S0-BPI-957

Structure, Culture, and Governance

Structure, Culture, and Governance

A Comparison of Norway and the United States

TOM CHRISTENSEN and B. GUY PETERS

JF
51
.C477
1999
west

ROWMAN & LITTLEFIELD PUBLISHERS, INC.
Lanham • Boulder • New York • Oxford

ROWMAN & LITTLEFIELD PUBLISHERS, INC.

Published in the United States of America
by Rowman & Littlefield Publishers, Inc.
4720 Boston Way, Lanham, Maryland 20706
www.rowmanlittlefield.com

12 Hid's Copse Road
Cumnor Hill, Oxford OX2 9JJ, England

Copyright © 1999 by Rowman & Littlefield Publishers, Inc.

All rights reserved. No part of this publication may be reproduced,stored in a retrieval system, or transmitted in any form or by any means, electronic, mechanical, photocopying, recording, or otherwise, without the prior permission of the publisher.

British Library Cataloguing in Publication Information Available

Library of Congress Cataloging-in-Publication Data

Christensen, Tom, 1949–
Structure, culture, and governance : a comparison of Norway and the United States / Tom Christensen and B. Guy Peters.
p. cm.
Includes bibliographical references and index.
ISBN 0-8476-9313-9 (cloth : alk. paper). — ISBN 0-8476-9314-7 (paper : alk. paper)
1. Comparative government. 2. Norway—Politics and government. 3. United States—Politics and government. I. Peters, B. Guy. II. Title.
JF51.C477 1999
306.2'09481—dc21 99-29374
CIP

Printed in the United States of America

The paper used in this publication meets the minimum requirements of American National Standard for Information Sciences—Permanence of Paper for Printed Library Materials, ANSI/NISO Z39.48–1992.

Contents

CHAPTER ONE

Norway, the United States, and Comparative Politics

THIS BOOK COMPARES the political-administrative systems in Norway and the United States, primarily on the central level, and the role that they play in the governing of these two countries. Attempting to compare these systems in their totality would be too great an enterprise for this one volume, but we will circumscribe the task somewhat by focusing on the capacity of these two political systems to supply a scarce commodity—governance—to their two societies. That dependent variable of "governance" is itself a rather broad concept, but we can explicate it by looking at the relative degrees of difficulty that these two countries have faced in performing the regulative and distributive tasks common to all developed democracies. Both have been able to govern, but they have faced very different challenges and have developed rather different means of coping with the demands of governance.

Our comparative task will be further narrowed by concentrating on two sets of independent variables—structure and culture—with a relative stronger emphasis on the first one. Again, this selection might not appear to narrow the task all that much, given that in some ways there is not much else that can explain the performance of political-administrative systems other than structure and culture. While that is true, the important part of our analysis will be demonstrating the way in which these two supposedly autonomous sources of explanation for social phenomena actually interact and reinforce each other, at least in these two systems that, each in its own way, have been successful in supplying governance to their economies and societies.

One obvious question about this enterprise would be why we would want to undertake a comparison of this sort with these two countries. On the one hand, it could be argued easily that since they are both stable, wealthy, democratic systems they are too much alike to make any useful and meaningful comparisons. In the language of comparative theory, these are "most similar systems" (Przeworski and Teune 1970), and therefore any comparison is likely to be less that fruitful. Why do we not adopt a "most different systems" research design and compare Norway with Niger, or the United States with Vanutu, if we really want to understand the differences among administrative behavior?

On the other hand, it could be argued that these systems are so different on so many dimensions that we cannot control enough variance to make the comparisons meaningful. Norway is unitary, parliamentary, consensual, multiparty and an advanced welfare state (Olsen 1983). The United States is federal, presidential, adversarial, two-party, and a welfare state laggard—and appearing to fall farther behind day by day (Jones 1995; Sundquist 1992). Given all these differences (and probably many more) the results of any comparison are not likely to be particularly meaningful. If there are important differences—as indeed there are—there are too many possible causes to be able to explain those differences efficiently or effectively.

It is, however, precisely because of this mixture of similarities and differences that we have chosen to do this comparison. These are two countries with a number of common features, yet also varying substantially in what their government do, in how they do those things, and in the "logics of appropriateness" in political life (March and Olsen 1989). If we look at the two countries from sufficiently far away, then they do appear very similar, but when we get closer we find interesting and important differences. Further, these differences appear to vary together to form coherent wholes, rather than just a set of variables varying at random. In other words, these patterns might be characterized as "syndromes," in much the same way that the symptoms of particular diseases vary together to enable to practitioner to detect the disease. The symptoms here are for the most part more benign indications, but they do appear to vary together.

Comparing Political-Administrative Systems

The main purpose of this study is to compare two political-administrative systems, United States and Norway. This means that we have to relate our study to an ongoing discussion about comparative method and analysis. The debate on comparative method seems, on a general level, to center around one major distinction; the methods of *difference* and *agreement*, a distinction developed first by John Stuart Mill (1872). This distinction has developed further into two different research strategies in comparative analysis, the most *similar systems design* and the *most different systems* design (Lijphart 1971, 1975; Przeworski and Teune

1970; Smelser 1973) being the phrases used conventionally to capture these alternative research strategies.

The *most similar systems* strategy is characterized by choosing to compare systems that are similar on as many variables as possible, except for the phenomena to be analyzed, the dependent variable (Lijphart 1971; Frendreis 1983). The systems should be markedly different on that dependent variable. The ideal is then to find as few independent variables as possible that vary among the cases and therefore can account for the variation in the dependent variable. The potential independent variables that are similar across systems are in this research design constants, that is, they cannot explain the variations in the dependent variable simply because they do not vary.

While the logic of the most similar systems design is appealing, it also presents a number of difficult methodological questions. For example, how similar is similar enough to include within a single category? Most social science variables are matters of degree rather than clear dichotomies so the research must make some judgments (Sartori 1991, 248–49). Further, are we sure that we have, in fact, identified all the variables on which the systems vary, or are there unmeasured factors that render the presumed causal explanation spurious? In some ways, therefore, the information requirements for this design are extremely high, and perhaps too high for the state of knowledge in comparative politics at the present time.

The *most different systems* strategy differs from the former design by selecting systems that do not differ on the dependent variable and then eliminating as causes all independent variables that differ among the systems (Przeworski and Teune 1970). This means that the ideal is to find as few as possible independent variables that are similar in different system and therefore can explain the similarity in the dependent variable. In this strategy one can establish a relation between an independent and a dependent variable in one country, and then test whether this relationship is also valid in other and quite different systems, that is, "micro replications" (Frendreis 1983; Rokkan 1966a;). If it holds across a range of different systems, then there is some certainty that the relationship is a valid one.

These two strategies have been discussed as if they were quite different, but Frendreis (1983) emphasizes that they are contained in the same type of logic. He argues that both strategies are directed towards maximizing the number of irrelevant independent variables for explaining the dependent variable(s). Or in other words: minimize the number of relevant and potent independent variables. If that number can be reduced to one, then some causal presumption can be established. The selection of particular design will have something to do with the knowledge about the dependent variable, either a priori or that gained during the research process. In statistical terms, these methods attempt to conform to the familiar litany of "maximize experimental variance, eliminate error variance, and control extraneous variance" (Kerlinger 1986).

One basic problem with both the similar and most different system strategies is that they have inherent limitations in generalizability. Frendreis (1983) therefore proposed a *mixed-system* research strategy that combines the two. In a mixed-system strategy the systems/cases will vary along both the independent and dependent variables, thereby allowing for a variety of comparisons to be made. This mixed design is, in essence, the type of research we are undertaking here. There are differences between the two countries, but there are also important similarities. Perhaps most importantly, there are fundamental conceptual similarities in the politics of the two countries. Therefore, we can use the same concepts for the two countries with some assurance that they have not been "stretched" to the point of clouding rather than facilitating the comparison (Bartolini 1993; Sartori 1970).

The comparison of different political-administrative systems is often done synchronically on a general, macro level, but it is possible to expand comparison in at least two ways. One is to analyze the development over time in the relations between central variables. With this model of comparison it is not only the national systems but also their characteristics at different points in time, as well as the dynamics by which they change, that are the subjects of analysis. Another comparative strategy is to combine international and intranational studies (Lijphart 1975), so that subnational units within a single regime are the subjects of analysis. This strategy has the advantage of holding constant (or nearly so) some of the potential confounding cultural variables.

How is our study design related to the distinctions mentioned above? Our design has some similarities with the most similar systems design in that we suppose a priori that United States and Norway are quite different on the dependent variable, even though each has been at least somewhat successful in governing. But we are not especially preoccupied with characterizing the systems on as many similar, independent variables as possible, even though they are both wealthy and democratic countries for example. We also believe it is difficult and actually undesirable to find a single, crucial, independent variable that really explain the difference between the systems. That reductionism would only render trivial the many complex and important features of the two social and political systems.

Our design is similar to the most different system method in that we focus on variations on quite a few independent variables, but are not preoccupied with identifying any one decisive independent variable that explains similarity in the dependent variable. The design is therefore most like the mixed-system research strategy, or that part of it focusing on both variations in independent and dependent variables. As a minor part of the study we will attempt to relate our cases to the two different categories of systems to which they are supposed to belong, and in that respect both the most similar and most different systems designs are relevant.

Our study is both *synchronical* and *diachronical* in its method; that is, we both study contemporary relations between the variables and analyze their historical development (Frendreis 1983). And we try to connect intrasystemic differences to intersystemic variations (Linz and de Miguel 1966; Rokkan 1966a). Both countries have marked regional variations that help to explain how the task of governance must be played out in order to be successful. We will not in any systematic way compare the United States and Norway to other countries in their categories of political-administrative systems, but we will use some data from other countries where it helps to contrast between our two countries, for example concerning some of the cultural variables.

Another way to characterize this research is as the search for mid-level theory (LaPalombara 1968). We are not interested in any grand theory that would explain everything about Norwegian and American politics. That would probably be a vain search anyway, and certainly would not be parsimonious. Rather, we are concentrating on some main aspects of the of the political-administrative systems on the central level—the executive, both political and administrative, and the legislative—and analyzing what the similarities and differences we detect there tell us about the capacity of the two systems to govern.

ANALYZING POLITICAL-ADMINISTRATIVE SYSTEMS

A Structural-Instrumental Perspective

There are many potential ways to analyze the function of political-administrative systems. Our point of departure is to focus on two sets of variables that traditionally have been principal perspectives in the literature of political science in general, and political theory more specifically. The first is the *structural-instrumental perspective.* This historically has been a central means of approaching political-administrative systems. An important argument, stated as clearly today as 2000 years ago, is that without a structural-institutional basis for a political system there is no political order and therefore no system (Sabine and Thorson 1973; Wolin 1960). Public activities have to be organized activities or they run the risk of becoming simply the exercise of power rather than authority.

We define here the theoretical point of departure for this perspective as a combination of political theory and organizational theory, and as some sort of "public organization theory" (Allison 1971; Gulick 1937; March and Olsen 1976; Mayntz and Scharpf 1975; Olsen 1983; Simon 1976). In this view, it would be inappropriate to separate political bodies and public administration from the fundamental fact that they are composed of organizations that have similarities to organizations found elsewhere. It would be equally inappropriate to separate these organizations from the fact that they are public, and therefore have very particular responsibilities, opportunities, and constraints (Walsh and Stewart 1992).

The formal structure of a political-administrative system is supposed to explain decisions taken in public organizations, the content of different types of public policies, and the actual outcomes of these policies within the society (Gulick 1937; Weber 1970). The argument is that the citizens are affected by the formal organizations of the political-administrative system from birth to the grave, more so now of course then some hundreds of years ago. For people working inside the political-administrative system the formal structure is of even greater importance. The structure selects and empowers potential decision participants, more clearly so the more unambiguous the structure (Simon 1976). Further, the formal structure also selects out different problems and solutions, and thereby shapes the content of public policy. This shaping could happen either indirectly, by privileging certain actors that carry certain definitions of problems or solutions, or more directly by specifying rather clearly the preferred outcomes of decision-making processes (March and Olsen 1976).

The formal structure of a system of this sort has two important elements, the organizational and the individual. This means that the formal, structural arrangements may be *interorganizational*, that is, regulating the relationships between different bodies within the policy system, and thereby can explain what different public organizations are doing. But the structure is also regulating the *intraorganizational* side of public bodies, for example by influencing the relation between different parts or subunits (Dexter 1990). And furthermore, formal structure creates different constraints on the individual political-administrative leader or any actor in the system, thereby channeling individual decision-making behavior in different substantive directions (Simon 1976).

The instrumental aspect of this analytical perspective is connected to the possibility for organizational design (Egeberg 1987; Sabine and Thorson 1973; Wolin 1960). The main argument here is that the conscious structuring of public bodies can fulfill certain public goals (Weimer 1995). The rationality of organizational design is an organizational rationality; that is, the rationality is related to the total, conscious organization of the different parts rather than partial and mechanistic (Scott 1992). This "holistic" manner of organizing is intended to modify the cognitive and other limitations of individual actors. Thus, we are concerned with the extent to which organizations provide frames of reference for their members that shape subsequent policy choices.

There are two conditions for effective organizational design: The political-administrative leaders must have the cognitive capacity to learn from their experiences with different organizational structures (Olsen and Peters 1996). Second, the leaders have to have the actual control or power to decide upon and implement new formal structures, not only formal authority, that of course is a precondition or potential (Dahl and Lindblom 1953). The ideal in this perspective is for organizational leaders to score high both on cognitive capacity and political control. In other words, those leaders must know what they want and have the capacity to bring it about.

A Cultural-Institutional Perspective

The cultural-institutional perspective argues, on the other hand, that the formal structure of political-administrative system is of only limited importance to understand the working of the system. The development of the administrative system is first of all understood as a slow and gradual adaption to internal and external conditions or pressure (Selznick 1957). This natural process of adaption creates an unique blend of normative values and norms, thereby giving the system a "soul" or culture that differentiates it from other social systems. This process of adaption can be seen as inculcating the technical, instrumental part of a system with values and norms, thereby giving it institutional features. The argument here is that different institutional norms and values develop into different decision-making behaviors, public policies, citizens, leaders, and societies (March and Olsen 1989).

The process of institutionalization focuses upon the integrative aspects of political-administrative systems (March and Olsen 1989). It "defends" the system by obstructing sudden and dramatic changes, for example, those that might be created by conscious design. Institutionalization is meant to be a stabilizing factor in a changing world. It is connected more to the survival and development of a social system than to instrumentally fulfilling any particular public policy goals. The citizens and members of public institutions have a feeling of common history, obligations, and reason (March and Olsen 1989). This set of norms and meanings will be a means for interpreting external changes and guiding organizational adaptations to them. The continuous creation of meaning, the different types of civic education, and the development of citizenship are all typical for a political-administrative system according to such a perspective. People are socialized into the institutional factors; that is, they internalize them (Selznick 1957).

The fundamental institutional norms and values in a political-administrative system may be of a general nature, emphasizing that the system must pay attention simultaneously to many different considerations (Egeberg 1987). These norms and values focus, however, more specifically on the importance of "the voices of the past and future," individual rights and obligations, "political areas of freedom," and so on—all considerations that have to be defended against more pragmatic and instrumental considerations than institutional leaders tend to push forward (Egeberg 1987). Institutional values, therefore, provide the interpretative framework for the instrumental actions that its members must undertake on a day-to-day basis, whether to maintain the status quo or to adapt to a changing environment.

THE DYNAMICS BETWEEN STRUCTURAL AND CULTURAL EXPLANATIONS

To understand the functioning of political-administrative system and the potential influence of structural and cultural variables, one has also to focus the relationship between these factors. One way of reasoning, and the one we will stress, is to expect consistency and reinforcement of the factors. This means that the operation and effects of structural and cultural variables are very nearly the same. Seen from this perspective, Norway is supposed to be both structurally and culturally homogeneous, while the United States is both structurally and culturally heterogeneous on a macrosystemic level. As noted above, these factors tend to operate as "syndromes" rather than as isolated factors, in part because they have evolved together over time to be consistent with one another.

One example of this manner of reasoning is to suppose that the principal political institutions in the United States—the presidency and Congress—are fragmented (or divided) when seen from a macro perspective and are also fragmented internally (fragmented administrative agency structures and fragmented committees in Congress), creating a fragmented culture with informal negotiations in "iron triangles" (with the interest groups).[1] These informal structures again result in fragmentation on a systemic level, albeit with substantial integration on a sectoral level (Freeman 1985). The same iron triangles, for example, function equally effectively when a new political regime coming into office, and can supply the talent needed to fill vacant offices in a bureaucracy that depends upon partisan appointments.

Norway, on the other hand, is usually characterized as having less fragmentation on the macro level, through the parliamentary system, less internal structural fragmentation, and a culture less characterized by disintegration, even though Norway also has some iron-triangle characteristics. Even in those cases of the close contacts between interest groups and government in Norway, there are structures and norms that force countervailing influences to be heard in the policy process. Likewise, although the administrative system in Norway is fragmented into a number of semiautonomous agencies, the political, legal, and cultural norms of the system serve to make it function in a more integrated fashion than would be true for the United States.

A more general version of this argument is that some kind of mutual adjustment characterizes the relation between structural and cultural factors

(Lindblom 1977). On the one hand, the conscious development and design of political-administrative systems is constrained by cultural variables. An example of this is that majority rule is constrained by institutional values and norms even in democratic settings (Egeberg 1990). Institutional processes build norms that began with a social character into the formal structures (Selznick 1957). On the other hand, the cultural-institutional development of political-administrative systems are constrained to a certain extent by conscious efforts to alter operational norms and values by manipulating rules. The same type of argument is relevant for more general decision-making behavior where institutional designs —for example, balanced-budget amendments or term limits—are used to develop (and if necessary enforce) agreements about behavior.

There is a different argument based on a belief in inconsistency and conflict in social systems (Simmel 1964). The logic of structural and cultural factors are different both in developing political-administrative systems and channeling the decision-making behavior of participants. Cultural norms and values will therefore undermine or modify the effects of structural factors. Similarly, structural factors may be modified using "tamper-proof" legal rules in order to force certain patterns of behavior as well as certain types of policy decisions (McCubbins and Schwartz 1987). These are used, for example, to limit the possibilities of "drift" in administrative organizations having the right to issue secondary legislation (Shepsle 1988).

A way of combining these argument is to suppose that the political culture of a country could have both elements of integration and disintegration. For example, the United States can have a culture that is sufficiently flexible and well-integrated that it allows a certain high level of fragmentation, while on the other hand the main effects of the culture are supposed to reinforce the fragmentation and centrifugal forces in the political-administrative system. The institutional fragmentation of the system provides almost all the contending social forces with an appropriate, friendly "target" for its political activities, whether that target is a Congressional committee, an administrative agency, or the courts.

A last argument is that the impact of cultural variables is dependent upon the type of formal structure in a political-administrative system. If the formal structure is clearly defined and the goals unambiguous, the effects of cultural factors will be minor (Selznick 1957). But if a political-administrative system has a relatively loosely defined structure and goals, the freedom for, and impact of, cultural variables may be substantial. The effects of the cultural variables can both reinforce and modify the effects of the structural factors. This way of thinking is some kind of general version of the major direction in the contemporary literature on organizational culture (Ott 1989; Peters and Waterman 1982).

The Central Variables in the Study

The two major sets of independent variables in this study are *structural* and *cultural* variables. We will emphasize the structural variables more than the cultural ones, a choice partly reflecting that the empirical basis in different, relevant studies is heavily biased in favor of the structural variables. Studies of political-administrative culture have been of a more general nature, not so much connected to more specific institutions, a feature leading us to treat these variables partly as a type of "frame" variables.

Interorganizational Variables

The structural variables are of two types: *interorganizational* variables and *intra-organizational* variables. First, the interorganizational variables contain several factors that shape the overall patterns of governance. They give insight into how the formal relationship between political and administrative bodies is organized; that is, how functions and authority is allocated. If we refer to the theoretical work on institutions by Weaver and Rockman (1993), these factors occur somewhat further up their implicit "funnel" of causality than do some of the other variables, but yet do have a pronounced impact on the capacity of a system to govern itself effectively.

The interorganizational variables can be divided into variables describing vertical and horizontal specialization, respectively (Christensen 1997a; Christensen and Egeberg 1997a; Egeberg 1987 and 1989b). Vertical, interorganizational specialization describes how political and administrative functions and authority is allocated across hierarchical levels, for example, between governmental departments/ministries and agencies or across geographical levels. Horizontal, interorganizational specialization focuses on how functions and authority are allocated between units on the same hierarchical level—that is, between political bodies or administrative sectors.

Accordingly, one relevant interorganizational variable is type of macro-organization between central, political-administrative bodies. The two systems we are analyzing are related to two different types of political macro-organization: a presidential type and a parliamentary type (Weaver and Rockman 1993). We will analyze whether the obvious differences in the macro-organizational structure of the two systems are important for decision-making behavior and governance. We will also elaborate on the two categories and try to differentiate them into other variables that can further specify the impact of macro-organizational structure. This variable is first of all one reflection horizontal, inter-organizational specialization, but has also vertical elements—for example, concerning special prerogatives of the executive. We will concentrate our analysis on the executive and the legislature in the two countries, while the judiciary will be treated rather briefly.

Two other interorganizational variables will only be mentioned, but not given any separate and thorough treatment. One is the principal characteristics of the geographical macro-organization in the two countries. The United States is characterized by a federal system, while Norway is more unified and centralized. The literature in this area discusses both to central political-administrative control systems and regional/local autonomy and the access of subnational actors to the central level. Although the United States is federal, for example, there is substantial, and increasing, interplay between the two levels of government, and policy is in reality increasingly a cooperative enterprise. This variable is primarily a vertical, interorganizational variable. The other variable is the corporative, macro-organization of the systems, reflecting how the central interest groups are integrated into the political-administrative system (Olsen 1983). The interest groups can be connected to the central political-administrative bodies in different ways—that is, through different organizational forms—and emphasize different institutions, thereby influencing the governance capabilities. This is a variable of a horizontal character, because this interorganizational relationship is basically of a collegial type, but it has also vertical features.

Intraorganizational Variables

These variables have also vertical and horizontal aspects. Vertical, intraorganizational specialization means how functions and authority formally are allocated across hierarchical levels internally in the political and administrative institutions on the central level. On the one hand, we have institutions that are highly centralized; on the other, public organizations that have allocated and delegated a good deal of authority and functions to lower levels. Horizontal, intraorganizational specialization shows how specialized functions and tasks are between units and positions on the same hierarchical level. Systems can be very specialized and fragmented, but also containing units and positions that have some more broad and general functions.

The first intraorganization concern based on these two types of variables is the central, lawmaking body in each of the two systems: the Congress in the United States and the Storting (Parliament) in Norway. We will both analyze the importance of the internal structures, horizontally and vertically, of these bodies, and also the main features of their demography. The same types of variables will be used in analyzing the executives—cabinet or president—and the central administration in the two countries.

We have pointed out the extent to which the interplay of these organizations helps define the policy-making system, but each of the institutions is important in its own right. This implies that the analysis of the central, political administrative structures and their impact on decision-making behavior and governance will discuss three interrelated sets of questions: (1) the internal structuring and function of the political-administrative institutions, (2) the interorganizational connections between the units of the systems, and (3) the relationship between the intra- and interorganizational features of the institutions.

Cultural Variables

Culture is generally thought to be more difficult to define and measure than formal structure. There are also few studies that more specifically analyze the culture that characterizes the relationship between institutions and actors in the central, political-administrative system. This leads to a search for cultural variables that could be indicators of relevant cultural features and connected to the main structural variables in the system. In other words, it is more difficult to find and discuss cultural variables that are "close" to the decision-making behavior and governance in the systems than to measure and analyze structural variables in this respect.

Cultural variables characterizing political-administrative systems may be divided into two parts: the more general and the more specific ones. The general type of cultural variables may function as a frame of reference for the more specific ones. It can be argued that different levels of culture—the systemic level, the political culture, organizational or institutional culture, policy-area culture, and so on—all influence governance. In this study we will focus on specific cultural variables, which are connected to the structural variables, but first outline some general cultural factors. Compared to the structural ones we will have more cultural variables, but have far fewer studies to lean on and therefore discuss them more briefly.

One systemic, cultural dimension useful for comparison is the number and type of cleavages in the society. One can suppose that the more divided and pluralistic a society is, the more the political-administrative system will also be heterogeneous and pluralistic. The two countries we are studying seem to be quite different on this dimension; the United States is a large country characterized by a number of deep and cross-cutting cleavages, while Norway is a small country and relatively more culturally homogeneous in this respect. This is not to say that there are not cleavages in Norway, but only that the political solutions for those cleavages may be easier to identify and to implement than in the United States, as will be discussed below. The other systemic, cultural dimension discussed is the one related to the cultural theory of Mary Douglas (1986, 1990). She combines two variables—*group* (people's affiliation to particular social groups) and *grid* (the extent to which behavior is constrained by rules)—to characterize culture in different systems.

We discuss three political-culture variables. The first variable is the attitudes towards legitimacy in a political-administrative system. On the one hand we have systems that are characterized by some kind of substantive policy ideal (i.e., that it is important for the public sector and organizations to "do the right thing," however that is defined). On the other hand we can identify systems that emphasize procedures—"rules of the game." In these systems it is important that public decisions are made through procedures that are reassuring for different actors. We will discuss whether the United States is typical of the former type and Norway of the latter, and the eventual implications of each.

The second variable is the attitudes towards law, judicial expertise, and due processes in the political-administrative system. The United States has a long cultural tradition (of course also in part structurally related) of the importance of legal due process in public decision-making, while in Norway this is rather unimportant. This variable must be relevant for analyzing the conditions for political control, through focusing the relationship between different powers inside the system, and the possibilities for actors outside the system to influence decisions.

A third variable is "trust in government"; that is, some kind of measurement of diffuse support for the system (Easton 1965). Is it the case that people in the United States respect their political and administrative institutions less than individual actors in the political-administrative system (Rohr 1987), and quite the opposite in Norway? Is Norway characterized by a general high level of trust, while people in the United States prefer local authorities with clientelistic features over the more remote federal bodies (ACIR 1994)? Are those federal bodies perceived as being more controlling of the public, even if they themselves often believe that they have little real influence over society?

The third set of cultural variables, and the one closest to decision-making behavior and governance, concern the specific, cultural variables related to the concept of *roles*—both institutional and individual. Institutional roles focus on the inter- and intraorganizational aspects of public decision-making. What characterizes, in general terms, the cultural traditions and contemporary role-enactment that are connected with more-or-less formal definitions of the relations between and inside different public institutions? Do these norms reinforce or modify the formal structural characteristics of the system? This enumeration of role relationships among institutions would include a number of different patterns of relationship within the governments.

First, we will focus on the relationships among the central political and administrative institutions within the systems—primarily the executive and law-making bodies. Is the cultural, interinstitutional tradition in the United States more characterized by separation of powers, conflict, and a formalized tug-of-war than in Norway? What elements characterize the informal relations between the bodies? Have they developed patterns of cooperation that are capable of overcoming the formal separation and conflict built into the structures? Are institutional interactions in Norway more or less conflictual than might be expected by looking only at the formal structural arrangements? What characterize the development of these institutions internally, and are these features reinforcing the interinstitutional culture? What characterizes the role of the central bureaucracy in relation to the other, more political institutions on the central level of government? Is the division between politics and administration more ambiguous in the United States than in Norway, given the long tradition of thinking about governing in this manner (Campbell and Peters 1988; Wilson 1887)? Is the relative importance of central bureaucracy higher in Norway than in the United States? Is the central bureaucracy in the United States more preoccupied with building alliances and relating to a large number of different premises for decision than in Norway?

Empirically, the analysis of this variable is somewhat special and divided into two. First, we discuss briefly the variable connected to the analysis of the structural variables in chapters 2 through 5. We will define the kinds of inter- and intrainstitutional culture that seem to have developed in relation to the structural development and its effects. Second, we discuss the variable more generally in chapter 6.

The second institutional role to be discussed is the culture of public management. This variable has been attempting to tap into some main features of public reform during the last decades. Is it because the so-called New Public Management involves cultural norms and values that are welcomed in the United States, but are not very compatible with the cultural tradition of the civil service in Norway?

Individual roles, both political and bureaucratic, are the third type we will focus on as cultural variables. Individual actors are acting on behalf of public organizations, and it is important to focus on their role perceptions and role constraints. First, we will emphasize the possible consistency in individual roles. Is this consistency lower in the United States than in Norway because of higher fragmentation and more considerations to satisfy?

Second, the exposure of and definition of accountability for politicians and bureaucrats may vary between the two political-administrative systems. Is it the case that politicians and bureaucrats in the United States are much more exposed to public accountability and held more personally responsible for decisions and outcomes than in Norway? Is the Norwegian bureaucracy characterized by less-exposed actors and more collective responsibility, or are there also important elements of personal responsibility? Which system appears to work better, or is its success dependent upon its acceptability in the particular culture within which it is implemented?

TABLE 1.1
STRUCTURAL AND CULTURAL VARIABLES

Structural Variables	*Cultural Variables*
Interorganizational variables:	*General, systemic variables:*
Political macro-organization—the executive and the legislature	Number and type of cleavage
	Group-grid-dimensions
Intraorganizational variables:	*Political-cultural variables:*
Vertical specialization in the executive and the legislature	Attitudes towards legitimacy
Horizontal specialization in the executive and the legislature	Attitudes towards law, judicial expertise, and due processes.
	Trust in government
	Specific variables:
	Inter- and intrainstitutional culture on central level
	Culture of public management
	Individual, political, and administrative roles: —degree of consistency —exposure and definition of individual accountability —individual discretion

Third, the *discretionary* influence on decision-making by public bureaucrats may be enacted in different ways in the two countries. Is it the case that the discretionary autonomy is more narrow and the political premises more important in the United States than in Norway, where formal autonomy is higher and professional considerations play a more important part in discretionary behavior? Furthermore, over the span of time, has the discretion that is available to civil servants in the United States and Norway been reduced or increased?

Table 1.1 sums up the independent, structural, and cultural variables that will be used in the analysis of the central political-administrative systems.

The Dependent Variable

Finally, the dependent variable in this study is political-administrative decision-making behavior. This variable actually can be divided into two types: First, there is behavior connected with the change of the system—internal and indirect variables, like behavior in reform, reorganization, or recruitment processes for the organization itself. Second, there is behavior directed towards the environment—external or output variables, like casework and regulation in different policy areas. It is often difficult to measure organizational decision-making behavior, so we must rely on focusing attention on central, individual actors acting on behalf of public organizations and observing how they interact with one another and with their society. Again, the interpretation of the manner in which these officials function is a cultural exercise, dependent on internal social norms for evaluation.

We use the dependent variables mentioned above to explore and illuminate a more general one—governance, the ability to govern the systems—as our ultimate or "real" dependent variable. That is, decision-making behavior is conceptualized as a factor that can be manipulated—by politicians and civil servants—in order to produce the more amorphous quantity of governance (Kooiman 1993). They can manipulate decision-making by manipulating the structures within which it occurs, or the patterns of organizational culture that inform decision-making. On the surface it is quite simple to change the institutional structure of governing. We have been arguing, however, that doing this without also considering any requisite changes in the cultural underpinnings of those structures may be of little real use. To govern effectively requires an integration of body and mind, structure and culture.

Summary

This study of politics and administration in the United States and Norway will focus on the capacity of those two systems to govern their economies and societies. Governance is difficult to define in any simple, unambiguous manner, but the basic phenomenon with which we are concerned is the capacity of a government to either control behavior directly, or create a set of incentives and disincentives for behavior, that enables it to produce desired societal outcomes (March and Olsen 1995). Phrased somewhat differently, governance is the application of the instruments of collective action such as authority and law to generate appropriate outcomes. There is a tendency to think that the best (or even only) way to create effective governance is through a highly centralized, coordinated political system characteristic of unitary and parliamentary regimes. Although the governing system of the United States often appears chaotic and/or stagnated, it can be seen to govern. Further, the contemporary reforms of governments in most industrialized democracies are moving away from centralized to more loosely coupled systems.

The capacity of the fragmented and deeply divided institutions and political culture of the United States to produce a pattern of government that many, if not most, Americans can accept points to the importance of cultural interpretations and explanations. The same formal patterns of governance would probably be a disaster in Norway, and certainly would be interpreted as such. The cultural variables are essential to understanding how these governments actually function. They are themselves, however, not sufficient but require the structural backbone of government for a complete understanding. Thus, this book will not be an exercise in discussing competing explanations but rather will be a demonstration of complementary and reinforcing roles of these two sets of explanatory variables.

NOTE

1. The "iron-triangle" conception is often argued to be outdated, but it is still (in our view) a good place to begin to understand American policy making.

CHAPTER TWO

Macropolitical Design of Government: The Legislative and Executive Related

THE MAIN THEORETICAL-STRUCTURAL ARGUMENT stated in chapter 1 is that structure really matters: that is, the way political-administrative systems are structured, either inter- or intraorganizationally, has substantial effects on the content of the policies decided upon. This means that the structural frames both constrain central political and administrative actors, but also provide them opportunities of control and influence over society and over other political actors. On the macro level, the design of the whole political system, the principal structural feature, seems to be stable or changing only slowly over a long period of time in most regimes. Politicians have more opportunities on a micro level, however, for fulfilling their goals by choosing certain organizational structures connected with certain policy outcomes and effects (Gulick 1937).

This chapter begins from the macro level and asks whether there are any important formal, structural distinctions among political systems, especially concerning the relationship between the legislature and the executive, that relate to the analysis of similarities and differences in governance in the United States and Norway. Lijphart (1984, 67; see also Shugart and Carey 1992) offers such concepts by stating that the most important distinction in legislative–executive relations in democratic regimes is that between parliamentary and presidential government. This chapter starts by discussing different theoretical perspectives on this distinction. The next part focuses on the history of the legislative–executive relationship in the United States as a presidential system and in Norway as a parliamentary system, ending with a contrast between the contemporary structure in the two systems.

This chapter is designed to serve several purposes in explaining the differences in executive and legislative relations in these two countries: First, to give an impression of whether the parliamentary–presidential distinction is analytically clear enough and empirical relevant to the two systems we are studying. Second, to give a background for analyzing the microstructures central to the performance of the two systems—that is, the internal organization of the legislature, the executive, and the central administrative apparatus. Are these structures closely connected to the overall distinction and what do we know about the effects of them? And third, to provide a background for the later discussion of how structure and culture are related, and thereby help to illuminate the crucial question of how governance is explained by these factors.

Presidential and Parliamentary Political Systems

The United States belongs to the category of presidential systems, and Norway to a parliamentary system. But does this make any difference—that is, is this distinction meaningful in analyzing differences in decision-making and governance between political-administrative systems? Are macrostructural independent variables of central significance to political-administrative influence and control?

One simple answer to this question is "yes," which is an answer based on two premises: The first is that macrostructural design *really* matters; the way the relationship between different political powers is formally structured has consequences for decision-making processes and governance (Egeberg 1987; Gulick 1937; Lijphart 1984, 1989). This argument can be supported by a central thesis in the *theory of bounded rationality* (a theory increasingly used in the study of public organizations): Decision-makers, like central political and administrative actors, have to make a selection of decision-making premises to cope with the complexity and problems of an often inadequate capacity for full rationality (March and Simon 1958). The formal structure surrounding them, which both constrains and provides opportunities, makes or conditions this selection and channels decision-making behavior towards certain directions (Egeberg 1987; Simon 1976). Stated more broadly, political actors in the executive and legislative branches make decisions to act based on the possibilities and constraints in their formal roles, defined by the *interstructural* (e.g., the formal relation to other powers) or *intrastructural* (e.g., the internal structure of their own institution) design.

The second argument is that presidential and parliamentary systems markedly differ in at least one fundamental, macrostructural respect: the constitutional role of the legislative and executive powers and the formal relationship between them (Lijphart 1984, 68; Weaver and Rockman 1993, 12). In presidential systems the legislative and executive powers are designed and constituted separately, as part of a separation of powers system (Fisher 1991, 6–12;

Sundquist 1988). They have a formal, independent constitutional status and separate constituencies, being based on different electoral systems and rules (Thurber 1991, 4). The citizens or electorate have two different agents or representatives, the president often designed to be executive and the legislature primarily a representative body (Shugart and Carey 1992, 1–7). Such a separation of powers system is often believed to a be reflection of heterogeneity in the political-administrative system, and results simultaneously in both fragmentation and flexibility (Polsby 1990, 19).

These differences illustrate that it is crucial to focus the historical-institutional background and development of such systems, meaning also to understand the cultural constraints on structural design. And that the choice of constitutional design both have advantages and disadvantages (Weaver and Rockman 1993). The choice of a political-administrative structure on the one hand provides opportunities to fulfill important goals and considerations in the political system, while on the other representing a "mobilization of bias" (i.e., organizing out of consideration of other values and effects), factors that may be important in other systems (Lijphart 1984, 74–78; Schattschneider 1960).

A parliamentary system, on the other hand, is characterized by a major modification of the system of separation of powers: it is formally organized and has a close connection between the legislative and executive powers. These powers still have different roles, but the formal basis of the executive power is the majority in Parliament, not a separate, independent basis, as in presidential systems (Hernes and Nergaard 1990, 85–86; Lijphart 1984, 68). This means that in a parliamentary system, as contrasted with the presidential one, the executive and legislative powers have the same constituency, namely, the selection of the cabinet is derived from the result of the elections to Parliament. The Parliament is the foremost representative or agent for the electorate, and the executive is more a representative for the Parliament, a "derived" sovereign authority (Shugart and Carey 1992, 1). Traditionally, parliamentary systems are said to originate from greater homogeneity in a political system and result in more centralized coordination and stronger political control, a characteristic believed to be both an advantage and disadvantage in a democratic society (Lijphart 1984).

To state this central argument more clearly, the difference between presidential and parliamentary systems is based on a combination of two factors: the degree of formal separation of powers and the electoral basis of different powers. Presidential systems have a formal separation of powers system that gives them a stronger potential for the legislative and executive powers to exercise a separate, actual-power basis in decision-making processes. This feature is reinforced by separate elections for president and the legislative branch. Parliamentary systems have modified the separation of powers system through coupling the basis of the executive power to the elections to the Parliament.

Another and much more complicated and diffuse answer to the question of the importance of the presidential–parliamentary dichotomy is that *it depends* on many other factors and conditions as to whether these main types of systems produce differences in governance. Weaver and Rockman (1993, 25) present an analytical scheme called a "two-tier model of determinants of government policy-making capabilities," and we will outline some of its main features here. The core of this model is that presidential and parliamentary systems differ according to a set of factors, but there are also variations within these two systems, thereby making the two types less distinct.

According to the Weaver and Rockman, presidential systems generally score low on party discipline in the legislative branch, on recruiting executive leaders from the legislative branch, on the degree of centralization of legislative power in the cabinet and on the degree of centralization of accountability. Parliamentary systems have quite the opposite values on all four variables, and these four factors are argued to have a significant influence on the decisions made in the systems of government.

There are, however, variations in the two main systems—the second tier of explanations. One way to differentiate between the two is to focus on *regime types*, or "the modal pattern of government formation" (Weaver and Rockman 1993, 19). Especially in parliamentary systems, there are some main regime types, labeled "multiparty coalition," "single-party-dominant," and "party government" (Westminster type) (Pempel 1990; Strom 1990). The differences among these types are also important for explaining differences in system behavior.

Inside each regime type there may be several different government types (Weaver and Rockman 1993, 22). These variations among governments are related to the effects of electoral rules and the rules and norms of government formation. Presidential systems can be divided into *unified* and *divided* systems, according to whether the party of the president is the same as the one holding majority status in the legislature. A multiparty coalition can be either a minority single-party government or an oversized coalition or a majority single-party government.

Considered from our point of view, the scheme presented by Weaver and Rockman is as a whole far too complicated for our purposes, and more generally for getting a clear impression of the two types of systems. It still, however, contains many central variables that will be used to characterize and analyze the political-administrative systems in the United States and Norway. We firmly believe that there are some major differences between presidential and parliamentary systems—as macrostructures, the microstructures derived from the macro, and the working of the systems—and that contrasting two such systems will show this quite clearly.

We will not go into many details in discussing how representative the United States and Norway are for their respective categories of regimes, but use some important factors in the analytical models as a frame of reference. Lijphart (1984, 68) argues that the United States is the only "pure" presidential system, while many others—France, for example—are more mixed types. Norway is typical for a northern European parliamentary government, especially if one disregards the federal elements in some such systems, but is quite different from the majoritarian aspects and party structures of the Westminster model, which in these respects are more similar to the United States (Lijphart 1989, 35).

THE LEGISLATIVE AND EXECUTIVE POWERS: SOME HISTORICAL TRENDS

We will now proceed to discuss some historical patterns of development in the two political systems. These will of necessity be rather brief descriptions of very complex developments. Rather than being full accounts of the complexities, they are rather intended to demonstrate the patterns of development and show the historical continuities in these systems, beginning with the United States.

Development Trends in the Relationship Between Congress and the President

The Constitution. Over 200 years ago the United States established a separation-of-powers doctrine together with a check-and-balance system, two complementary structural principles (Fisher 1991, 10). The framers of the Constitution feared despotism and domination of one of the political powers, and therefore created a complex system of checks and balances, of cooperation and conflict, both between and inside the several branches of government (Rockman 1990, 1–3; Sundquist 1992, 22–29).

The checks for the Congress in the Constitution against the other branches, besides its strong position in legislative processes, include approving executive appointments, war and treaties, impeachment powers, and an important role in the amendment processes, among other things (Sundquist 1992, 36–45). The checks within Congress consisted of two chambers, modeled after the existing state legislatures. The chambers perform many of the same functions, but are rather loosely coupled. They had two different electoral systems, encompassing different electorate, selectorates, and the length of terms. The Senate, originally elected by the state legislatures, could potentially limit the excesses of the directly, popularly elected House. Bicameralism was important for reaching agreement about the Constitution. The construction of the Senate favored the small states and the House the larger ones.

The Presidency. There was a substantial discussion among the framers about the presidency, and it was not obvious that the executive power should be single, not collective (Berman 1987, 25). Unambiguous responsibility was weighed against fear of a monarchy-like executive. In the Virginia Plan it was proposed that the president should be elected by the legislature, because one thought that the electorate was not sufficiently informed to choose the electors (Berman 1987, 23; Sundquist 1992, 29–31). Many states were in favor of a single-term president, and they discussed different length of terms, before they decided upon two four-year terms (not written into the Constitution, but practice following Washington). The final design of the presidency underlined, however, checks against the Congress by giving the president an independent electoral basis, and veto power in legislative processes to counteract Congress attempting to undermine his power.

The president is elected through an electoral college, thereby removing the president formally, but not actually, from direct popular control (Kernell 1991, 94). This electoral college system is itself in a way balanced, combining a proportional rule concerning the electors, biased somewhat in favor of the smaller states, with a "winner-takes-all" system favoring the large, urban states and thereby traditionally also liberal or moderate presidential candidates.

The judiciary also got an central instrument for self-protection, the power to declare laws unconstitutional. That power was not explicitly stated in the Constitution but was derived from the "supremacy clause," making laws and treaties made in pursuance of the Constitution the "supreme law of land." Following from that, someone had to decide which laws met that criterion and that was a self-proclaimed right of the judiciary.

Development Trends. Originally the House was seen as the most representative institution in the central political system in the United States. But with the gradual broadening of the electorate in the first decades of the last century, the Senate and presidency were also seen as representative bodies of electorate, organized on a different territorial basis (Maidment and McGrew 1991, 97).

The development since then seems to have had four main elements: First, history generally shows that periods of strong presidents have been followed by periods with Congress increasing its power (McCubbins 1991, 115). Second, Congress from the beginning had executive functions to perform, beside its obvious legislative. After the first five presidents, the Congress began to design the executive branch in detail, and the first hundred years were characterized by relatively weak presidents with small staffs, with administrative authority constrained by delegation from Congress (Sundquist 1987, 261). The last part of this period in particular was characterized by Congress and political parties, together with states (coupled to the Senate) and local government, as the principal actors in American government (Rockman 1990, 11).

This period laid the foundation of the close connections between some administrative units and the committees in Congress (Arnold 1986, 9–11). But it was increasingly problematic for Congress to administer public policies, leading to pressure to delegate authority to the president and his professional administrative staff. This has lead to the next trend, in this century—the power balance between the branches has gradually changed in favor of the president (Berman 1987, 3; Rockman 1990, 4–5). To put it in Sundquist's words (1987, 262): "The President has developed to be the general manager of United States, while Congress is the board, with substantial political power." The formal constitutional power division has remained the same, while the president through a variety of laws has been delegated substantial autonomous authority, but with great reluctance and ambivalence in Congress. The contemporary conflicts between the new Republican majority in Congress is reviving some of the older conflicts.

Fourth, after World War II the Congress and the modern presidency have intensified continuously their elaboration of mutual instruments of control, making the political system much more complex and fragmented. Congress both allowed the president to build up a stronger apparatus, but on the other hand counteracted the president by creating its own counter-bureaucracies in Congress (Polsby 1990, 15–17). Through different reforms, the widening of its statutory powers, enhanced legislative oversight, and so on, the Congress in the 1970s tried to check the president. But this trend also created additional fragmentation, conflicts in Congress, and a greater workload.

The 1980s was characterized mainly by a president actively using his available instruments of influence showing that the maintenance, or strengthening, of the power of the president has to do with more than his constitutional power, namely, his unique political position and potential for taking issues directly to the voters to obtain public support (Mezey 1985). So the long-range trend of a gradually stronger president does not seem to have been changed by the Congress trying to bounce back in the 1970s, and again in 1995, after the Republican victory in the Congress elections. The United States has developed a stronger presidency, but not necessarily a weaker Congress (Polsby 1990, 15).

Mutual Means of Control. Most of the mutual means of control have a long history. The framers gave the Congress the power to establish offices, the president the power to nominate, and the Senate the power to advise and consent to the nominations, even for the appointment of the president's own cabinet (Fisher 1991, 23, 36). The Constitution forbids members of Congress to hold federal offices during their period of office, thereby potentially preventing corruption and manipulation by the president, but also preventing anything approaching a parliamentary system.

But the Constitution did not foresee development of political parties and the gradually increasing of workload placed on the legislature, factors leading to delegation of authority from Congress to the president, development of more instruments of mutual control, and greater conflict between the powers (Fisher

1991, 117). The constitutional framework as originally articulated was appropriate for the late eighteenth century but not really well adapted to the pace of the late twentieth century.

The delegation of authority from the Congress to the president can be seen either as abdication or maintenance of the congressional power (McCubbins 1991, 116). One the one hand, delegation can actually remove influence from Congress and make it difficult to control the president and the executive administration. On the other hand, delegation may be effective when seen from Congress, since Congress can get involved with the executive branch on a broad basis and use many indirect and direct means of controlling the delegated authority (McCubbins 1991, 116–17). Delegation historically has been related to three central and interconnecting areas in the relationship between the Congress and presidency: the legislative process in general, the budget process, and reorganization authority.

The power of the purse has always been an important instrument for the Congress, but used both in pork-barrel processes, increasing the budget, and to constrain economic initiatives from the president and the working of the executive agencies in a detailed way (Fisher 1991, 134, 136, 186, 283). Already in the 1860s the committees in Congress had established a veto-power constraining executive agencies and their programs. The Budget and Accounting Act of 1921, creating the Budget Bureau (later OMB [Office of Management and Budget]) in the White House, is seen as an important structural change that increased the influence of the president in budget-decision processes (McCubbins 1991, 117). And the Congress tried later to counteract by creating the Congress Budget Office through the Congress Budget and Impoundment Act of 1974 (Sundquist 1992, 281).

Congress developed the legislative veto as an instrument of control from the 1930s onward, prompted by increased presidential activity in legislative processes, and as a result of increased discretion of the executive officials. And during the same decade legislative processes delegated more authority to the president concerning reorganizations, but at the same time developed a one-chamber, later a two-chamber, veto procedure.

Arnold (1986, 4–5) suggests that comprehensive reorganization plans in this century have been important instruments for most presidents in gaining political influence. What he labels the "Managerial Presidency" has had, without too many guidelines from the Constitution, to organize unorganized administrative elements into a bureaucratic apparatus—a problem in a fragmented system and with the control from the Congress. The Managerial Presidency implied trying to strengthen the organization of the budget process, the staff of the president, and the policy-analytical capacity in the executive. The first president to have a strong program for administrative development was Theodore Roosevelt.

The picture of a relatively stronger presidency, based on an increasing resource base, institutionalization, and delegation from Congress, can however be elaborated. The president's authority over many large components of the executive apparatus is more formal then actual, since the authority is subdelegated in many policy areas and far away from the reach of the president (Fisher 1991, 288). This again is furthering the development of subgovernment and alliances between committees/subcommittees in Congress, interest groups, and agencies.

Changing Political Composition. Historically, the relative political composition of the legislature and the executive has changed significantly. In the last century and up to after World War II, unified government was the rule (i.e., a divided electoral system produced systematically more often than not a majority for the same party in Congress and presidency). In the period from 1947 through 1992, however, divided government has been dominant, compromising three-quarters of the time period (Cox and Kernell 1991, 3). Interestingly, a large proportion of the state governments in the United States have also been divided during the same period.

During the last 120 years the Republicans have held the presidency for 72 of them, having a clear dominance. The Democrats have held a majority in both chambers of Congress for 56 years during this period, the Republicans for 38 years, and 26 years have had a divided Congress. The Republican presidents therefore historically often have faced a divided or hostile Congress, and more so after World War II, while the presidents from the Democratic Party most often have governed with a politically compatible Congress.

The impact of these divisions must, however, be related to the strength of partisanship—that is, has the party division been of major significance, or has cooperation been experienced across party lines? The answer to this seems to be that partisanship, and therefore divided government, has been of greatest importance during the later decades. Also, the Democratic Party itself often has been divided between its Northern and Southern wings.

Ticket-splitting, either between the president and Congress or inside the Congress itself, seems to occur in about half of the presidential-election periods. This has been explained, but not empirically proven in a convincing manner, by the voters' tendency, on the one hand, to choose conservative, Republican presidents who favor tight fiscal control, and on the other hand by having money and resources flow to groups or constituencies by voting for the congressional Democrats (Cox and Kernell 1991, 3). Scholars seem to disagree about the effects of divided government on the content of the public policies in different areas and effective government, but agree more about institutional developments and the use of means of control.

Divided government, especially in the postwar period, seems to foster more presidential control of the bureaucracy, more administrative capacity in the

Executive Office of the President, more political appointees, and more frequent use of the presidential veto. And the dominant party in Congress—the Democrats—has tried to obstruct Republican presidents by reducing their discretion in formulating and implementing policies and controlling their bureaucracy through primarily legislative means. The executive bureaucracy especially has been a battlefield between the president and an opposing Congress, resulting in the accumulation of numerous means of control, primarily based in Congress (Cox and Kernell 1991, 6; Gormley 1989).

Especially in the last two decades, the following congressional development trends have been related to divided government (Mayhew 1991, 191): (1) Congress is increasingly engaged in "micromanagement" of the executive branch; (2) Congress has increased the staff controlling the bureaucracy and increased its oversight hearings; (3) Congress is using more legislative vetoes than before; (4) Congress has decided upon more detailed rules for constraining bureaucratic discretion; and (5) Congress has tried more consistently to restrain the power of the president—for example, through the War Powers and Congressional Budget Acts. All these changes are disputed by critics of the relevance and effects of divided government, and we will discuss their arguments later.

Trends in the Relationship Between the Storting and cabinet in Norway

The Constitution. The Norwegian Constitution of 1814, based on a union with Sweden, was built on Montesquieu's constitutional principles about the division and balancing of powers. It imitated both those ideas, reimported to Europe from the United States, and the English political system, but also was designed along Norwegian traditions (Hernes and Nergaard 1990, 17). The Constitution established three independent powers: the executive, the legislative, and the judiciary, the latter defined with a weak potential-power position. Some of the founders wanted a stronger institutional basis for the judiciary, like in the United States, but the result was more in accordance with Montesquieu's original thoughts about this power.

The Constitution stated that the king, the leader of the executive power, should choose his own cabinet, which originally meant that it, as in the United States, was to be as adviser to the leader of the executive (Hernes and Nergaard 1990, 39). The king was situated in Stockholm and the majority of the cabinet and all the ministries were in Christiania (later named Oslo). There were actually two prime ministers: one Norwegian Prime Minister in Christiania (after 1873) and one in Stockholm, heading the division there. Two ministers alternated between the two divisions of the cabinet; the Stockholm division was working as a branch or delegation (Heidar 1983, 18).

The Constitution in Norway originally defined a stronger executive than in the United States, both in its own right and concerning shared functions. The king had the right to initiate laws and could postpone laws for up to six years. This was a personal veto, since the Constitution did not define a cabinet (Hernes and Nergaard 1990, 43). The king also had the right to appoint senior civil servants and his formal military power was strong.

The *Storting* was established with two chambers, the *Lagting* (one-quarter of the representatives) and the *Odelsting* (three-quarters), mainly for practical reasons in the legislative processes, an organizational construction of little political significance and not connected to different constituencies. This was also a reflection of the fact that Norway had no real nobility that could dominate a first chamber in coalition with the king, as originally thought by Montesquieu.

Between 1814 and 1884 Norway had a system of "moderate popular sovereignty" (Hernes and Nergaard 1990, 30). The legislature was elected through an electoral college (until 1905) by the people, but that meant only a very small proportion of the population. For example, in 1859 about 8 percent of the population enjoyed the right to vote, and 31 percent of the enfranchised actually voted (Kaartvedt 1964, 146). Norway gained universal suffrage in 1913, but men had obtained their voting rights in 1889. The Storting was elected for three years, and until 1869 met only once during that period. The judiciary was partly self-recruited, partly chosen by the king, and the executive was based on a combination of inheritance and choice by the monarch.

The macroconstitutional design was first of all characterized by a strong monarch who dominated his cabinet and was often opposed by the Storting (Kaartvedt 1964, 242–43). The conflicts and tensions in the system were exaggerated by the fact that Norway was the inferior part in a union headed by a Swedish king. This intercountry conflict was background for another perspective on this postconstitutional period, namely, to view the system as some kind of "civil servants state" (Hernes and Nergaard 1990, 42, 58; Seip 1963). Many members of the cabinet and Storting had the same social background and job experience as the civil servants in the central administration. This similarity created a potential of shared attitudes and interests, blurring the constitutional separation and cleavage between the executive and legislative powers, giving a potential of weakening both the union and the constitutional macrodesign.

The Norwegian Constitution was not quite clear about the distribution of functions between the powers; it was especially ambiguous about the fiscal and budgetary power of the Storting. Unlike the U.S. Constitution, the Norwegian Constitution defined few checks-and-balances procedures between the powers. The king could, as mentioned, initiate lawmaking processes and postpone their implementation through a personal veto (Hernes and Nergaard 1990, 46). The Storting had the power of the purse, implying that it could control the money for administrative positions that the executive wanted to establish. The Storting also could decide upon detailed budgets that bound the executive. But since the Storting convened so seldom, these instruments of control were, for a long period of time, not very actively employed.

The Judiciary. The judiciary developed over time a certain scrutiny function as to whether laws were made in accordance with the Constitution or administrative practice was in accordance with law. This was to many an imitation of the role of the courts in the United States, which together with Norway are the two countries with the longest-lasting constitutions and scrutiny systems (Smith 1993, 31–32, 189).[1] But the judiciary has never had any politically active role in Norway, unlike in the United States. One reason for this is the weaker formal position of the courts in the Constitution, a position that was made even weaker after 1884 and the parliamentary principle. Second, the political focus on the choice of the members of the Supreme Court has never been the same in Norway as in the United States, where every such choice is a political battleground. Third, the judiciary in Norway has had much less freedom in selecting what types of cases to deliberate upon. And fourth, the homogenous and centralized system in Norway has had less potential of conflict than the federal system in the United States.

One feature of the system in Norway that illustrates that the judiciary has some power is that civil servants were both formally and actually nearly impossible to discharge. This is a reflection of the judiciary balancing the power of the executive and the legislature.

The Period 1814–1884. Taken together, one can conclude that the checks and balances between the powers in the Norwegian political-administrative system in this period between 1814 and 1884 were rather limited and biased in favor of the king. The Constitution was also somewhat ambiguous or used somewhat different principles concerning the creation of separate careers for the different powers, an important point for Montesquieu. Civil servants in the central administration could not, after 1814, be elected to the Storting (Hernes and Nergaard 1990, 50, 60, 62). The representatives of the Storting could be chosen as ministers, but had to give up their seats in that institution for the remainder of the election period and had no direct access to the deliberations taking place there. The members of the Supreme Court could, however, be members of the Storting at the same time, as they also did to a certain extent in this period, but the principle appears first of all to illustrate the weak political position of the judiciary and did not result in any political dominance of the courts.

During the period of 1814 to 1884 the king was building his power on the division of power principles and could not, as in many other countries, obtain any support from a conservative first chamber dominated by an aristocracy. Seen from the perspective of popular sovereignty, it was trying to curb the power of the king by two ways: First, the king had to be brought under the control of the cabinet, and second, he had to be brought under the control of the Storting (Hernes and Nergaard 1990, 43). In the decades leading up to 1884, the Storting, now influenced by the mobilization of the peasants, tried to obstruct the king in many major issues and thereby undermined his power. The Storting succeeded, for example, in organizing some major parts of the central administration—the directorates—as quasi-independent bodies, outside the ministries (Christensen 1997a).

The Breakthrough of the Parliamentary Principle. In 1884, after vigorous conflicts, a long standoff, and an impeachment process, the traditional division-of-power principle was replaced by a parliamentary principle. This principle united the power in the "parliamentary chain"; it meant the dissolving of the political importance of the king in Norway.[2] It was a breakthrough for popular rule, an extended popular sovereignty, based on a coalition between peasants and urban intelligentsia (Olsen 1983, 43). After 1884, the Storting election decided the composition of the cabinet and more generally heavily influenced the recruitment of the executive political leadership. Majority rule coincided with other major changes, such as the establishing of the first two parties (Conservatives and Liberals), the suspension of the royal veto, and allowing members of cabinet to enter the Storting to participate in debates and answer questions (Hernes and Nergaard 1990, 85).

One problem with establishing the parliamentary principle was that once the cabinet was established, it was potentially no longer very dependent upon or closely related to the debate-and-decision activities in the Storting. The actual power could be moved from the Storting and to the cabinet, or especially to the parties and party groups in the Storting (Hernes and Nergaard 1990, 89). But the Storting had, and still has, some countervailing means to moderate this potential problem. One means is the principle of ministerial responsibility, which makes the cabinet and its members accountable or responsible for how their administration performs its functions and tasks (Hernes and Nergaard 1990, 92–93). This principle is connected to the demands on ministers to produce relevant and correct information to the Storting (stated in a law in 1932), and violations of this principle have always been politically serious.

Ministerial responsibility puts pressure on ministers, both to secure the loyalty of civil servants and in relation to the scrutiny from the Storting to which they are exposed. The principle means, however, not that the Storting has some kind of direct formal instructional authority towards the administration, as in the relation between Congress and some central administrative bodies. It is a more indirect instrument of control, but still powerful.

Another instrument of parliamentary control is that political sentiments in the Storting change during a mandate period, and the majority may, for political reasons, lose trust in the cabinet. One problem with this is that the cabinet can put pressure on the Storting and demand a vote of confidence, or threaten such a vote. If the Storting then is not prepared for changes it can loose influence and prestige.

A third instrument of a more general character is that the voters, if they do not like what the cabinet is doing, can vote for parties outside the cabinet. More generally, this means that the relationship between the Storting and the government is dependent upon the political composition of the Storting. During the period 1945 to 1961 the Labor party had a majority in the Storting, and the importance of the party and party group, together with the cabinet, was dominant. In the period between the wars and during the last two decades, Norway has had changing coalitions and many minority governments, leading to a stronger Storting in general and relatively less influence by parties and cabinet.

Taken together, changes in government in Norway have had different causes in different periods: in the period from 1884 to 1905 these changes were mostly caused by elections; from 1905 to 1940, change more often resulted from demanding and not getting a vote of confidence; and after World War II, mixed causes such as the parties' changing leadership, illness, conflicts, and disintegration of coalitions generated change (Hernes and Nergaard 1990, 87). Even though since 1884 the parliamentary principle has resulted in a close connection between the Storting and cabinet, Norway has retained a substantial division of power and functions between the powers. Contributing to this is that the Storting and the government do not intervene in the work of the judiciary, and the members of the cabinet temporarily give up their seats in the Storting.

Mutual Means of Control. The close connection between the executive and legislative powers defined in the parliamentary principle is, however, biased in certain ways in that it favors the Storting. The Storting has different means of control over the government: institutional, legislative, and economic. The cabinet must discuss certain issues with the extended committee of foreign affairs in the Storting. The government also informs the Storting through reports and proposals, and the representatives can ask ministers to clarify an issue in the standing committees in the Storting (Hernes and Nergaard 1990, 92).

The cabinet in Norway always has an important advantage in having an administrative apparatus with expertise to its disposal, a factor that has been even more substantial because the Storting traditionally has been reluctant to build up its own administrative resources. Both tendencies are quite opposite to those found in the United States. The cabinet uses this apparatus for preparing decisions for the Storting and thereby heavily influence the decision premises. And the control over the implementation of the decisions in the Storting gives government possibilities for influence, since more and more decisions, especially after World War II, are of a framework—as opposed to detailed—character.

But all this taken together does not imply that the cabinet dominates the Storting more than ever. The cabinet and its civil servants have a fine-tuned feeling for what is politically feasible in the Storting, engage in "sounding out" over a lot of issues, and are constantly adjusting to the political sentiments in the Storting (Christensen 1991a). This is a general feature of the system, but the mechanism is more actively used in periods with minority governments or loose majority coalitions.

Ebb-and-Flow. The influence of the Storting over government in general and the cabinet in particular can be seen in an ebb-and-flow perspective. In the period from 1814 to 1884 the Storting had relatively minor influence compared to the king and the civil servants, both in the cabinet and in the central administration. This was caused by the inferior position of Norway in the coalition with Sweden, by the constitutional macrodesign, and by the weak popular basis and political activity of the Storting. After the breakthrough of the parliamentary principle in 1884, the Storting gained substantial political power based on the new and close connection to the cabinet, but also on the establishment of political parties and mobilization of new groups into political life, broadening the popular basis of the legislature.

In the period after World War II there has been a tendency to conclude that the Storting is gradually losing political influence (Olsen 1983, 39–42; Rommetvedt, 1994). This is argued to be connected to: (1) a strong Labor Party dominating in longer periods and capitalizing on the potential in the parliamentary principle; (2) stronger resources of coordination connected to the cabinet; (3) increasing problems of capacity for the Storting, resulting in more general types of law and delegation of public authority; (4) a growth in resources and expertise in the central administration; and (5) the increasing importance of interest groups in close integrated participation with the public apparatus (Olsen 1983, 148–87).

One general argument against this conclusion of the decline of the Storting is that this political body has had, and still has, important symbolic functions besides the instrumental functions, and that it cannot be judged solely on an instrumental basis (Olsen 1983). Therefore it can be difficult to analyze and measure the real power of the Parliament. But some analysts have even pointed to a "reparliamentarian trend" in Norway during the last decades. This argument takes as a point of departure some kind of "hypothesis of contraction" (Jacobsen 1977): After a period of delegation of authority, allowing the administration and interest groups more participation and influence in public decision-making processes, the political leaders have wanted more political control again. This was furthered from the 1970s onward by economic stagnation and a "conservative political wave" in Norway. Of importance for this argument is also that Norway has experienced unstable parliamentary conditions the last two decades, thereby creating more interest in, and importance of, what is going on in the Storting. A last argument is that during the same period the Storting, relatively speaking, has increased its administrative resources substantially (Rommetvedt, 1994).

Comparing the Contemporary Macrostructure of the Systems

The overall and enduring impression of the political-administrative systems in United States and Norway is one of structural heterogeneity and homogeneity, respectively. First of all, the United States, as a presidential system, has a formally defined separation-of-powers system, dividing the power between the judicial, legislative, and executive powers. The system is also characterized by formally giving each of the coordinate branches an important role to play in the political system. The separation of powers is, however, not clear-cut but organized in such a way that each power can limit or balance the authority of the others, a system of separate powers sharing functions (Davidson 1988, 9, 14; Fisher 1991, 6–12; Mezey 1991, 14). The president can, for example, formally be involved in the legislative process through his veto power, or Congress in the executive process through appointments, investigations, reviews, legislative vetoes, and so on, thereby creating a potential both for conflict and cooperation (Fisher 1991, 93, 116–20, 134–35; Maidment and McGrew 1991, 40–42). This checks-and-balances system is, of course, motivated by not wanting to give any public institution or group of actors too much power and by a desire to slow the process of political change (Wilson 1987). It is also a reflection of societal heterogeneity and that the political system has to attend to many considerations simultaneously.

One structural feature in this presidential system is the variation between unified and divided government, that is, changing according to the party of the president and the majority in Congress. Even more complicated, it is combined with divided party control over the Senate and the House of Representatives (Thurber 1991). Divided government has been the rule rather than the exception since World War II in the United States (Oleszek 1991, 91). A quite common conclusion about the effects of divided government is that it is strengthening some of the main effects of the structural separation-of-powers system, namely, less-effective political coordination and control, more ambiguity concerning accountability, unpredictability, more political conflict, and less importance of political parties, thus creating a systemic paralysis (Burns 1956; Cox and Kernell 1991, 4; Destler 1985; Fiorina 1989, 1992; Sundquist 1988, 1992). But divided government can also, under certain conditions, foster compromise and competition in initiating legislative changes (Berman 1987; Oleszek 1991).

Some scholars have emphasized, however, and we think rightly so, that the difference between divided and unified government is not always that significant; for example, in concerning the lawmaking role of Congress and its investigations of the executive branch (Mayhew 1989, 1991). One can generally argue that this distinction does not add very much to the formal roles and functions of the executive and the legislature. This latter view has, however, started a scholarly debate about criteria for evaluating performance and nonperformance in a divided government system (Sundquist 1992, 102–6).

In a way, Norway has also an underlying separation-of-powers system, but since 1884 it has been very much modified and has become quite different from that of the United States. The parliamentary principle formally organizes a close connection between the legislature and the executive power, a prerequisite of homogeneity, thereby forcing the two sets of actors to cooperate (Hernes and Nergaard 1990). The power of the executive, weaker in this system than in the presidential system in the United States, is derived from and based on the legislative power—not on a separate constituency as in the United States. The actual distribution of influence between the two, however, is varying and primarily dependent upon the changing composition of the Storting. Generally, the relative influence of the executive power, and the party, is its highest when a single party has the majority in the Storting (Pempel 1990; Strom 1990). This was the situation in Norway with the Labor Party from World War II until 1961 (Nordby 1993). This period has been labeled "the one-party state," a concept indicating problems for democracy, but also political effectiveness (Seip 1963).

On the other hand, the relative importance of the Storting is probably at its greatest when there is some more political instability and ambiguity, resulting from changing majorities and coalitions (Hernes and Nergaard 1990; Olsen 1983). This has generally been the situation in Norway during the last 20 years (Heidar and Berntzen 1993, 55–56). Currently, the situation is somewhere between these two extremes. Norway has a Labor government, governing alone in a minority position, seeking support from issue to issue from different parties. But the nonsocialist parties, having the majority, are unable to agree upon forming a cabinet, a feature strengthening the Labor government.

The difference between a majority and minority government in a parliamentary system in not that large. A minority government still has to have support from the majority of the Storting, can attract support from different parties in different policy areas, and as such have a relatively stable political basis. More parties can have an influence on the policies, the arena for negotiations is often the Storting, and this can be seen as some kind of democratic safeguard and balancing of the power of the executive.

Of importance in Norway is also that the system is formally far more centralized than in the United States, both because of the parliamentary principle and the formally very weak political role of the judiciary. This feature adds to the overall homogeneity of the Norwegian system. Concluding this point, it is worth emphasizing that the United States has a more formal division of powers (with separate power bases and elections) than Norway, with its parliamentary connection of the executive and legislature. The system in the United States is blended with a principle of separate powers sharing functions, while the Norwegian system has a similar principle but to a much lesser degree.

A second difference between the United States and Norway concerns the importance of political parties. The parliamentary system formally implies a party-based executive power (Weaver and Rockman 1993, 8). This means, as in Norway, that the potential for developing strong parties and party discipline is high. The dominant way for a group to gain political power or for an individual to make a political career is through the collective action of the parties (Heidar 1988).

A presidential system is, through the structural separation of powers, much more open to the influence of different actors and potentially less based on parties and party discipline. There are, as in the U.S. system, many formal ways to exert political influence or to have a political career, and the system is more based on the fragmented, individual effort, demanding vast resources, than the collective capacity of parties. The system of primaries, for example, exposes politicians to the voters and makes them more individual political entrepreneurs. According to this method of reasoning, a natural conclusion is that the formal constitutional macrostructure is both a precondition for, and influenced by, the party structure.

Third, the degree of macrostructural homogeneity may also be connected to the degree of overall internal structural homogeneity of the different powers in the political system, as will be discussed in the following chapters. A common feature of the political system in the United States, discussed in the following chapters, seems to be a structural heterogeneity inside the executive and legislative powers especially. We will suppose that the more internal the heterogeneity, the more fragmented the system, the more potential there is for a complex pattern of contacts and interdependence both between different bodies inside the public apparatus and in their relationship to the task environment. In Norway, the structural homogeneity both inside the legislature and the executive power seems to be much higher, probably reinforcing the overall homogeneity of the system.

Fourth, the difference in structural homogeneity between the political systems is also shown in their geopolitical systems. The United States has a deeply rooted federal system, leaving relatively substantial political influence to the different states, while Norway has a much more homogeneous geographical system, characterized by relative centralization. This sometimes inhibits national policy-making in the United States but also permits substantial experimentation and flexibility (Kettl 1984).

Fifth, an important feature of political systems is how they organize their relationship to interest groups. In the United States, in accordance with the structural heterogeneity in the public apparatus, the system is widely open to the influence of interest groups. Interest groups have developed a complex pattern of contacts to the executive and especially the legislature, and their style is often ad hoc, single-issue oriented, and rather aggressive (Weaver and Rockman 1993, 27– 28). In the Norwegian system, as in many parliamentary systems, the participation of interest groups is more desired by the government, more regulated, organized, and on a more continuous basis, often called "integrated organizational participation in government" (Olsen 1983). Their formal contacts are mainly limited to the central administrative apparatus, and participation is much more dominated by hierarchy, peak organizations, and a strong membership base.

Is it possible, based on these overall structural characteristics of the two political systems—a heterogeneous and a homogeneous one, respectively—to have some preliminary thoughts about governance, about the main features of the decision-making processes and behavior they are resulting in? One way to assess the systems is to focus on their decision-making effectiveness, on their ability to reach collectively defined goals and implement them (Egeberg 1997). Another way is to focus on how easy it is to identify political and administrative responsibility or accountability within the systems (Weaver and Rockman 1993, 15–16). A third way to judge is according to the systems' responsiveness: their ability to reflect or represent on a broad basis the political sentiments and interests in the population (Olsen 1983, 43–45; 1988a, 157–60). These are all important aspects of how political elites legitimatize their activities within political systems.

In the United States, the separation-of-powers system is normally criticized for scoring quite low on decision-making effectiveness and identification of responsibility (Sundquist 1988). The system is said to be divided, with many different centers of power in conflict with one another and blaming one another for any failures. There are diffuse lines of authority and patterns of responsibility, with inconsistent and unstable policies, and with avoidance of difficult decisions (Weaver and Rockman 1993, 1–2, 16–17).

But a presidential system may allow more open democratic control of its activities through many points of access and potential influence for outside actors than a parliamentary government does (Wilson 1987). And it may be more open to access from outside groups, more effective and flexible to produce decisions that are satisfying to different groups, and overall more innovative. Another way to put this is to say that the system scores relatively high on responsiveness. The openness, structural heterogeneity, and complexity of the system may, however, be systematically biased in favor of actors and groups having ample financial resources.

A parliamentary system such as the Norwegian one is said to score higher on decision-making effectiveness and accountability, because the concentration of power and strong parties makes the priorities, political control, and identification of responsibility easier (Weaver and Rockman 1993, 16–17). But such a system can also be criticized as being too centralized and regulated and threatening to people's individual freedoms (Wilson 1987). In addition, too much central loading and sensitivity to political changes may generate ineffectiveness. The responsiveness of a parliamentary system may also be assessed differently: On the one hand the system may be seen as consistently furthering and aggregating the interests of people, but the centralization of power may, on the other hand, be related to a potential bias in favor of a limited political elite and strong interest groups, and an insulation from ordinary citizens (Olsen 1983, 13–14).

These preliminary thoughts about different aspects of governance will be developed and discussed in the following chapters, based on an intrastructural analysis of the legislatures and executives in the two countries.

CONCLUSIONS

The analysis of the developing and contemporary executive–legislative relationship first of all illustrated important differences: The constitutional starting point for the two systems had some real similarities. In the United States, the presidency was originally thought to be dependent on elections to the legislature, a possible parliamentary trait, but received a separate power base through separate elections, emphasizing the separation-of-powers system stated in its Constitution.

The Constitution of Norway in 1814 created a system of separation of powers, of "modified popular sovereignty," with a king as the executive leader, choosing his cabinet, with a judiciary either self-recruiting or selected by the executive, and a legislature based on a limited direct, popular basis (Hernes and Nergaard 1990, 30). The historical development of the Norwegian state has retained some of that separation, while strengthening greatly the position of elective officials vis-à-vis the monarchy.

But these originally relative similar systems developed in different directions. The presidency in the United States was formally granted relatively weak powers in its Constitution, while in practice the Constitution has developed a strong presidency, especially during this century. Paradoxically, very few scholars for that reason have concluded that Congress is a politically weak body—quite the contrary. One reason for this is, of course, the system of checks and balances, and sharing functions.

The Norwegian Constitution created a strong executive headed by the king and a relatively weak legislature, the Storting. Since 1884, the Storting has however become a strong political body (with certain progressive and regressive periods) without developing a weak executive, which is a reflection of the parliamentary principle, a feature that underscores the much more homogeneous character of the Norwegian system. The two powers (i.e., the executive and legislative) are much more cooperative partners than in the United States. Hernes and Nergaard (1990) label this relationship in Norway between the executive and legislative as "power integration," which reflects how the constitutional close, formal connection between the two has created a strong informal relationship.

In the United States, the separation-of-power system primarily reflects and produces heterogeneity and thereby conflicts, with the result that there have been few factors binding the system together. This separation is exacerbated by the weak party system in the United States, the independent judiciary, and the multiple cleavages in American political life.

NOTES

1. Norway has in this respect been much less like other Western European countries than the United States. The U.S. court system in Norway has been both a possible ideal and something to avoid, especially the excesses of the system.

2. Since 1884 the king primarily has had ceremonial and symbolic functions. Interesting enough, after dissolving the union with Sweden 1905, Norway decided in a referendum to choose a monarchy instead of a presidency, and elected a Danish prince to be king (Haakon VII).

CHAPTER THREE

Legislatures: Historical Trends and Contemporary Structure and Decision-Making Processes

WE HAVE OUTLINED SOME main features, historical and contemporary, in the relationship between the executive and legislative powers in the United States and Norway, showing that these systems are quite different in their macropolitical structure. The potential for these systems to perform well concerning different aspect of governance, like political responsiveness, accountability, responsibility, and effectiveness, also have briefly been pointed out. These differences illustrate a central theorem in organizational studies: That organizing (here meaning macro- or interstructural design) creates a bias towards certain considerations, while other considerations become less important (Gulick 1937; Hood and Jackson 1991; Schattschneider 1960). Organizational structures are good and bad at doing certain things at certain periods of time, but cannot meet all preferred goals and interests all the time.

Chapter 3 focuses the internal structure of the legislatures in the two countries, both historically and today. The central questions to be answered are: What are the main characteristics of the internal organizations of the legislatures and how have they developed? What are the political-administrative or decision-making significance of these structures? Are they systematically strengthening or modifying the effects of the different political macrostructures in the two systems? We will also briefly discuss the connection between the internal-structural features of the legislatures and their relations to other central actors in the systems, primarily the executive.

Development Trends in the Legislatures

Some General Tends in the U.S. Congress

Congress differs from parliaments in other Western countries by placing more weight on individual voters and constituencies and less on parties. The American political parties have gradually been further weakened through this century, with a possible strengthening again during the last decade. Under the Clinton administration the Democrats appear to have developed in a more disintegrated direction, while the Republicans have strengthened their internal party cohesion significantly over the past decade.

Congress has experienced a steadily increasing work load in this century, reflected in more issues being discussed and decided upon, more days in session, more committee hearings, among other things (Davidson 1992, 4–7). The domination of a strongly partisan Republican party during the mid-1990s enhances the capacity of Congress to process that work, but the fundamental division of the system still restricts its capacity to produce legislation.

In the era of the "conservative coalition," from the mid-1930s until mid-1960s, Congress was dominated by a small group of senior leaders in safe seats with a narrow policy and lawmaking agenda and enjoying bipartisan, conservative cooperation. This was a hostile environment for activist presidents and their supporters. The reform era, from the mid-1960s and to the late 1970s, is characterized by more liberal Democrats coming into Congress and increasing pressure from active, discontented interest groups, mediated by active presidents. These reforms have created a more egalitarian and open Congress, allowing broader participation and influence by members (Davidson 1992, 12). This is also a period of strengthening of the staff and other resources of more members of Congress, with closer contacts with constituencies in not-so-safe seats, and for many lawmaking activities, promoted by policy entrepreneurs. The postreform era in the Congress, a period of "contraction," is characterized by greater economic concerns and cutbacks, more omnibus bills, more hierarchy, more routines, and more partisan voting (Davidson 1992, 13).

The leadership in Congress has been relatively weak and divided in most of this century, but this last period produced relatively more hierarchy and a stronger leadership, especially among the Democrats. The trends of change are stronger in the House than in the Senate in this period. The postreform Congress appears to focus its attention on general interests and open decision-making processes, an effect related to more general lawmaking and change of culture, and more difficult times for articulating and influencing through special interests (Quirk 1992, 310–19). Under the Clinton administration, special interests are believed to have enhanced their influence again. The impact on the deliberative aspects of Congress are more mixed. Congress has on the one hand increased its capacity for handling information through establishing new structures, increasing resources, and expertise. But Congress is also experiencing greater external pressure and overload. The attention paid to negotiations among members (in committees and out) has also increased in the postreform Congress, but so too have partisan cleavages and conflicts.

The main internal structure of the Congress has not changed much over time, with the committee structure as the backbone, more so for the House than the Senate. The committee structure also creates a stable leadership in Congress over time. During this century the committee structure has become more differentiated, implying a more elaborate subcommittee structure (rationalized somewhat in the last decade) for coping with an increased workload. The staff and other resources for the committees are also gradually increasing. But the members are under increasing cross-pressure, caused by internal pressure, heterogeneous and active constituencies, and interest groups, and are forced to choose and to prioritize where to involve themselves in the decision-making processes (Hall 1993, 166). Generally, members are experiencing more problems of attention, and the mandates from their voters are more ambiguous and flexible (March and Olsen 1976, 1983, 1989).

The structuring of the Congress in this century, also connected to the reforms, has generally weakened its traditional deliberative character (Dodd 1993, 428). Congress has gradually been more complex and fragmented, characterized by special interests, strategic behavior, manipulation of procedures and debates, more technocratic and less collegial and party based (again with tendencies in the opposite direction for the party strength and partisanship lately). The decision structure in the Senate is traditionally more open, and it is easier to participate in a floor debate without belonging to the relevant committee (Hall 1993, 162–63; Smith 1992, 171). In the House, floor debates have been and remain more regulated and the importance of specific committee memberships and chairs is more tangible.

The most important reform period in Congress in this century took place during the 1970s. The reforms adopted at that time, mainly in the committee system, were aimed at weakening the dominance of seniority and the chairs in committees. The reason for these were partly to strengthen the subcommittees and to broaden the participation of all members, but also to strengthen other leadership positions—mainly the party leadership. Congress traditionally has not been a structure that can centralize political power, act quickly in politically important matters, or plan comprehensive policies (Sundquist 1981). Congress is characterized by individualism, and this feature has not changed substantially through different internal reforms. Some reforms, like those enacted during the 1970s, have made Congress even more fragmented—a tendency making it more difficult for the president to relate to Congress and thus fostering a larger presidential staff (Ornstein 1982).

Some Development Trends in the Norwegian Storting

The Parliament in Norway, the Storting, played an important role in writing and defending the Constitution of 1814, when the Danish–Norwegian state was dissolved and a new Swedish–Norwegian union was established (Olsen 1983, 42). During the last century, the Storting successfully defended national values and "lay governance" against "the civil servants' state" and Swedish interests. The establishment of an institutional identity in the Storting was an integral part of dissolving the long union with Denmark. The Storting was influenced also by the mobilization of new groups into politics, especially farmers during the 1860s and 1870s. The breakthrough of this group, together with the urban intelligentsia, made the parliamentary "revolution" possible and thereby also created the values connected to lay rule, the political amateur, and the small staff in the Storting.

Before 1884, the Storting played the role of a "shadow cabinet" and used the slogan "all power in the hall of the Storting" (Olsen 1983, 43). The Storting "revolted" against the king, the cabinet, and the civil servants, but the conflicts were not very strong and never violent, which was in accordance with the Norwegian consensus tradition. The establishment of the parliamentary principle in 1884 was the major breakthrough in this struggle, making the Storting a "real power" in the political system and potentially the foremost representative of the people, with the gradual widening of the election rights (Arter 1984, 17).

The second large mobilization in Norwegian politics, that of manual labor, came comparatively late—during the first decades of this century. The mobilization was through labor unions and the Labor Party—in principle still a revolutionary party in 1928. But from 1930 onward, the Labor Party has been a parliamentarian reform party. After many unstable governments between the wars, the parties made a "Joint Program" in 1945, establishing a long tradition of consensus in the postwar period. Since then, the Labor Party has been the largest and most dominant party in the Storting and has been in power for 39 of the last 52 years.

Since 1884, the Storting has primarily been based on the political parties, not directly on voters and constituencies. This is especially true since the 1920s, when Norway introduced proportional, parliamentary elections on the regional level based on party lists, replacing a system of majority elections in single-member constituencies. And in the same period, as an effect of the new election system, the number of parties in the Storting started to increase substantially. Since then, the typical Norwegian parliamentarian has always been less exposed to the citizens than the members of Congress, always seen more as an actor in a collective body than an individual politician. Voter turnout in Norway (like other

Nordic countries), has been substantially higher than in the United States, normally around 80–85 percent of all voters (no special registration required) (Arter 1984, 13).

The political parties are still relatively strong in the postwar period, especially as parliamentary parties. But the parties, and therefore also the Storting, have changed. The most obvious example of this is that the Labor Party gradually has become less ideological, less a mass party, with a more formalized and complex organization, with problems of retaining membership and loosening ties to the labor movement.

Just as in the U.S. Congress, the Storting has experienced a gradually increased workload in this century, occurring especially during the postwar period. More issues have been discussed and decided upon, and the Storting meets for more days and for longer hours (Olsen 1983, 45). The Storting has been coping with this increased load by leaning more on the work of the standing committees (relatively few, traditionally one for each ministry) and handling issues in a more general way (i.e., concentrating more on the framework of budgets and laws and less on details). This development is similar to that in the U.S. Congress. The Storting has also always been, and still is, an arena for small, symbolic local issues to be exposed in the name of geopolitical interests, reflected also in the fact that the representatives in the Storting are seated according to their county affiliation (mixing representatives from different parties).

The Storting has gradually become more formalized and hierarchical since 1884, especially after World War II. This means that the leaders in the Storting, primarily the leaders of the party caucuses but also leaders of the most powerful committees, control most decision-making processes and debates. They tightly control the timing of the issues, the exposure of conflicts, who says what in debates at certain times, and so on. This structuring is strongly modifying the deliberative aspects of the debates in the Storting, making innovation, surprises, and "discourse" rather infrequent (Olsen 1983).

The main internal structures of the Storting have not changed much over time, and substantially less than in the U.S. Congress. The Storting is a much less complex organization than Congress, has relatively fewer resources, and experienced no major reform periods after 1945, even though there were small changes during the last decade; namely, in the growth in administrative resources and some despecialization of the committee structure (Rommetvedt 1994). The backbone of the Storting is its dual character, based on the party caucuses and the committees, a structure that Congress was gradually more approximating during the 1980s, even though the Storting still has a more centralized power structure.

The Storting has, in contrast to Congress, always had rather few formal connections to actors in the environment such as interest organizations, except for the party organizations and cabinet (for the ruling party), because it does not want to be committed in the earlier stages of the decision-making process (Olsen 1983, 65–71). This means that representatives have been reluctant to participate in public committees, there has been few formal ties to the central administration, to interest groups, to individual citizens. But this does not mean that the Storting is isolated; quite the contrary, it has a very tight informal network of contacts to many groups and actors (Hernes and Nergaard, 1990).

Changing the Internal Structure in the U.S. Congress

The U.S. Congress is originally based on some main considerations: lawmaking, allocating resources, balancing the power of the president, and a representative institution for the interests of the electorate (especially the House of Representatives) (Maidment and McGrew 1991, 96). Wilson (1987) emphasized that it is something quite different to discuss the formal principles in the Constitution and the praxis that has evolved in relation to those principles. From his point of view, the most important part of the "unwritten" U.S. Constitution is the internal organization and procedures of the Congress, especially the committees.

The committees in Congress started to hire part-time, temporary clerks in the 1840s (Malbin 1980, 11). In 1856, the first committee in the Senate established permanent secretarial help, and before the end of the century all committees had some such help. In 1891, the House committee staff totaled 62, with 42 in the Senate. Some few decades into this century the staff was a combination of personal (started in 1893) and committee staff, but the staff was also gradually more politically oriented. In 1926, the first nonpartisan, committee staff emerged.

The Legislative Reorganization Act of 1946 formalized the professional staff element in Congress and started its substantial expansion. During the period between 1950 and 1980, the budget of Congress increased forty-five times, the committee staff in Congress increased eightfold, the personal staff fivefold, and supporting agencies, such as the General Accounting and Congress Budget Offices, increased in importance (Malbin 1980, 4). The increase in staff size from the 1950s onward was relatively more political in character again and more connected to strong chairs of the committees and the party to which they belonged. Through the Legislative Reorganization Act of 1974, the minority party in the committees also strengthened its staffs. This increase also generated a broader range of activities and probably a higher potential for conflict. Congress has also hired extra help for legislative activities and for contact with constituencies, and the staff of the top leadership positions also increased, together with the different ad hoc groups.[1] Taken together, however, the personal staff of the members are the most numerous element of congressional staffing.

There are many reasons for this staff expansion (Malbin 1980, 6). One important one is the need to develop specialized expertise in Congress, thereby increasing the quality of the decisions and creating more independence of the environment. This is also connected to two other reasons: influencing national political issues and improving the standing of Congress in the media. The administrative expansion is also related to coping with increased overload and the problems of capacity, but here the effects are more uncertain, since the staff also generates a lot of activities and ambiguous lines of authority.

Traditionally, the Senate, with its smaller size and longer periods in office, has placed more weight on floor debates—the deliberative aspect—than the House, underscoring its more national and media-exposed position (Smith 1992, 171). The Senate has been more open and internally democratic, but also more atomized or individualistic than the House (Smith 1993, 259). Also traditionally, the Senate has had weaker partisan cleavages than the House, party caucuses have had a weaker status, leaders have had weaker formal positions, and there has been closer cooperation between majority and minority leaders (Smith 1992, 179). But historically, the importance of committees and their chairs, based on seniority, also has been substantial. They have controlled committee procedures, set the agenda, and dominated the debates. Furthermore, this hierarchical element has been strengthened by an uneven distribution of staff and other resources.

The leadership positions in the Senate developed during the first three decades of this century: first the party caucus chairman, then the floor leaders (Smith 1993, 260). Since then, the majority- and minority-leader positions gradually have been more institutionalized. The reforms of the Senate committee system in the 1970s, after pressure from younger and more liberal members, created a more egalitarian Senate by weakening the seniority principle and spreading the resources more evenly among the senators (Ornstein et al. 1993, 20, 23). The reforms were also a reaction to an increased workload and demands from politicians to obtain important political positions. This was seen in the increase of the number of committees, and especially subcommittees. The staffs also increased—an attempt to curb this increase was only temporarily successful.

Taken together, this development has both advantages and disadvantages for senators. It reduced some of their load, but on the other hand also increased their problems of capacity by increasing the number of committees and subcommittees to which they must attend, and the staff has generated more activities (Ornstein et al. 1993, 33–34). The changes in the committee system have changed the decision-making processes in the Senate. It has weakened the traditional strongest actors, the chairs in the standing committees, moved the legislative battle to the floor debates, and the agenda-setting and legislative oversight to the subcommittees. The substantial increase of time in staff and resources for the senators has been coupled with a change in their task profile, from a more narrow and internally focused to a more extroverted one, placing relatively more weight on external activities such as reelection campaigns, fund-raising, media exposure, and so on (Ornstein et al. 1993, 20).

The type of leadership in the Senate has differed between the parties. The Democrats have had a more substantial potential for a strong leadership, fostered also by their long-time majority position. Their floor leadership during the post-war period consistently has been strong, centralized, personal, and informal (Ornstein et al. 1993, 23). Instruments of control have included the selection of committee members, party strategy, the timing of issues and decisions, etc. From 1988 onward, the leadership power has been somewhat more divided, but still relatively strong, with deep involvement in substantial policy questions, the use of party task forces, and revitalization of the party caucus (Smith 1993, 259). The Republicans have traditionally had another leadership style in the Senate: more formalized, institutionalized, and decentralized, although party leadership under Senator Bob Dole was more personalized.

Generally, leaders in the Senate are more powerful than before, but it has also developed a broader group of leaders sharing power (Smith 1993, 272–80). They have changed the rules and practice of debates and decisions, thereby furthering control and curbing obstruction, even though the filibuster is still available for use as an instrument for preventing action. They have increased the importance of the party organization and the resources connected to the leadership positions. And they are more active and policy-generating than before, especially the Democrats, before the Republicans gained the majority in the Senate again in 1995.

The most central, structural component of the House has always been the committees, an instrument first for handling a political body with many members, but later very important for handling an increasing workload. There is also a long tradition for a dominance of committee leaders, starting in the 1920s, with powerful, conservative chairs—known as tough negotiators and expanding the domains of their committees (Dodd and Oppenheimer 1993, 50). But an increasing share of Democrats in the House generated after a while an opposition to this system. By the 1950s and 1960s, there was a growing dissatisfaction among some young Democratic representatives about the organization of and power distribution in the House. They believed that the House was too hierarchical and closed. This group of representatives, together with a revitalization of the democratic caucus in the House, initiated the reforms in the House in the 1970s (Dodd and Oppenheimer 1993, 46).

The reforms developed in a somewhat different direction: They weakened on the one hand the powerful, Democratic committee leaders by implementing secret ballots for choosing committee leaders, increasing the number and power of the subcommittees, and making most committee meetings open. The increased differentiation in the committee system was also an answer to increased workload. More members in the House, with their expertise and political solutions, could participate in the decision-making processes, but this again has probably generated more activities, problems of capacity, and political polarization. The reforms increased the potential for more diffuse limits of responsibility and conflict over domains of each committee.

The power of the Speaker increased and one created a budget committee, both potentially integrative and coordinative changes, but also potentially creating new power centers. The Budget Act of 1974 tried to integrate the House and Senate through the process of reconciliation and by establishing joint budget and economic committees. This trend became more obvious in the 1980s, furthered during Ronald Reagan's presidency, and tighter spending controls. The committee leadership became more centralized and hierarchical again, especially in committees working with economic and budgetary questions (Dodd and Oppenheimer 1993, 49), and committees for different policy sectors declined in importance. All together, this weakened the influence of ordinary representatives and created more internal conflicts.

Since the 1970s, the power of the Speaker and the party leadership in the House became more important, especially for the Democrats. The Democratic Caucus also became more important for policy discussions. This changed a trend that had occurred during most of this century—namely, that the party leaders in the House had been weak and the institution more characterized by "committee government" and seniority (Dodd and Oppenheimer 1993, 54). The reforms of the 1970s created a stronger Speaker of the House, with more control over committee recruitment, more resources, and more control over the legislative and budget processes. But the actual success of the Speaker has, however, varied since then, depending largely on the holder of the position.

Another trend during the 1980s was the broadening of the House leadership by the Democrats. This change underscored the increasing importance of the majority leader and the whip, but created addition positions to share the power and democratize the leadership. This development seems to be the result of a long period of divided government, but also resulted from an ambivalence among party members concerning the autonomy of the leadership (Sinclair 1993, 237–38).

The Internal Development of the Storting

The main structure of the Storting has remained relatively unchanged from 1814, except for the establishing of parties and party groups in the Storting after 1884. After 1945 the Storting has gradually been modernized, both the formal structure and the processes of decision, through rationalization, specialization, standardization, and formalization. The Storting has always been reluctant to build up a large staff, a principle often thought to be in accordance with a layman-rule (Olsen 1983). Unlike many other Western legislatures, it has been politically difficult to argue for developing sources of counter-expertise in the Storting. This is also a reflection of some kind of a "passive" role definition, meaning that the Storting shall decide on laws and budget and not care so much about what is going on in the governmental apparatus, while the government shall prepare and implement the decisions. This role definition also reflects a deep consensus-orientation and homogeneity within the system.

In 1869, when the Storting first established an office with a manager, the Storting had only two persons on the staff, and the size of the staff increased very slowly for many years (Hoff 1964, 447; Olsen 1983, 55). But after 1945 and especially during the last two decades, the staff in the Storting, both technical and professional, has increased substantially, along with staff financed by the parties. In 1981, the size of the staff was 264, and in 1994, 387 (Rommetvedt 1994, 6). However, each representative has only 2–3 staff members, technical and professional—a comparatively small number. The professional staff in party caucuses and the professional advisers have increased the most, three to four times during the last two decades, but are still not more than one-fifth of all employees in the Storting.

Each Storting elects six presidents, a president and vice-president for the whole Storting, and the same for each of the two chambers, the *Odelsting* and the *Lagting*. The president of the Storting has a formal high status, next to the king (Olsen 1983, 62). But this position is in reality of modest political importance. The presidents of the Storting have many ceremonial functions, but of primary importance are practical tasks in organizing the internal decision-making processes, directing the debates, and attending to the "Rules of Procedure" (Arter 1984, 139, 141, 149). One central way of modernizing the Storting has been to change these procedures, but the Storting has had and still has time-consuming decision-making processes that have been handed down.

Ever since 1884, the party groups have been the backbone of the parliament. Since 1921 there have been fourteen different parties represented in the Storting, four parties for the whole time period. The party caucuses are organized with a chairman, vice-chairman, and a board and have grown organizationally more complex, with more representatives in the caucuses, more complex issues to discuss, and more complex patterns of contacts with the environment. The main task of the caucuses is coordination, both with the party and other relevant groups outside, and the activities inside the Storting (Olsen 1983, 63). The coordination is done in caucus meetings, where the group leaders and the representatives from relevant committees, according to the issues discussed, are important, eventually together with ministers from the party. The party caucuses are also very important for developing party identity and integration, for signaling the party profile, and for initiating implementation of the party programs.

Party caucus meetings have traditionally been characterized more by discussions, consensus, and "sounding-out," than by conflicts and voting. There is also a norm of party loyalty after the discussion of the issues in the caucuses: It is expected that the representatives vote in accordance with caucus decisions. Divergence from this norm is normally allowed in moral–religious questions and in some geopolitical issues. Traditionally, the Labor party has had stronger norms of loyalty than the other parties, which is also a reflection of its size and central political position.

The committees in the Storting have a long tradition, dating from the beginning in 1814. They have always been the main instrument of specialized attention in the Storting. The constitution says nothing about the committees, although the "Rules of Procedure" contain rules for their activities (Selle 1980, 1). The committees have traditionally had no formal authority, but have gradually become more important in preparing all issues in closed meetings and practically oriented discussions. Each representative is a member of one standing committee—normally only one—and there are no subcommittees, quite unlike the differentiated system in the U.S. Congress (Arter 1984, 161–77, 191).

Since 1824 the Storting has had standing committees, a system that replaced a large number of special committees. Over the years there has been some reorganization in the committee system, both mergers and splits, but the number of committees has been relatively stable—between twelve and nineteen. Since the 1930s the committees, like the structure of the ministries, have gradually grown more specialized (Selle 1980, 10). During the early 1990s there was some rationalization of the structure of committees, making them more general in profile, for example, by dissolving the committees of fisheries and agriculture. This change aimed at decreasing the special and segmented attention structure of committees and representatives and create more intersectoral capacity for coordination, but may also create more conflicts, problems of capacity, and ambiguity by connecting many committees to more than one ministry (Rommetvedt 1994, 32).

Over the years, the committees have grown more specialized in their internal organization. Originally, the committees had only a chairman and a secretary (traditionally and still a general role), and all the other members had few duties. But since 1890 the committees appointed a vice-chairman and special members to prepare certain issues or parts of them (Danielsen 1964, 161; Selle 1980, 4). This system has since then gradually been more elaborated, and increased the capacity of the committees substantially. But specialization in the committees, taken together with the growth in their administrative capacity, has created more workload and more conflicts (Arter 1984, 163; Rommetvedt 1994, 9–11).

There is a hierarchy of committees, with the committees on foreign affairs and finance at the top, but many representatives also want to remain on the lower ranked committees to which they are first assigned, because of their special interests and competence (Arter 1984, 169, 177; Hellevik 1969). This latter trend has been sharply decreasing after the last elections, partly reflecting a decrease in the share of representatives reelected and partly due to the party caucuses wanting to rotate their members (Rommetvedt 1994, 8). Seniority is less important than in many legislatures, since the party caucuses decide where to place members, often according to their status and rank in the party (Arter 1984, 200).

During the period between 1814 and 1900 there were approximately 100 representatives in the Storting; in 1919 that increased to 150, and now is at 165 members.[2] Since 1869 the Storting has had yearly sessions and from 1945 an obligatory fall session, but the Storting still does not meet from late June until early October (but it is now discussing starting a month earlier). This means historically that the enormous expansion of the agenda in the Storting has not been followed by a comparable increase in representatives and allocation of time. The members have got more and more problems of attention and capacity. This again indicates the importance of the factors that are influencing what type of political questions the representatives are attending to, often resulting from certain processes of political definition and interpretation (Olsen 1983).

After 1945, the number of issues discussed in the Storting has doubled. But the number of recommendations from the committees has decreased by one-third since the early 1970s (Rommetvedt 1994, 19). More important is that the recommendations demonstrate many more conflicts within the committees. For many years, 15 to 20 percent of the recommendations showed some disagreement; during the period between 1989 and 1993 the number was 59 percent.[3] "Fraction remarks" from the parties in the committees (i.e., comments showing deviant opinions) increased from 2,090 during 1977–1981 to 11,791 during 1989–1993. This does not mean that the Storting is ever more conflict-ridden, since many issues are solved by an unanimous vote (Arter 1984, 289). But the figures about conflict indicate many things: First, that the issues have become more complicated and fragmented, making it easier to frame minor questions of an important symbolic character. Since there is less money to use and less political willingness to use what money there is, this seems to foster more talk and "hypocrisy" than political action (Brunsson 1989). Second, a period of minority government means that the traditional socialist–nonsocialist division is not that clear any longer (Arter 1984, 287–89). The Labor Party, for example, must lean on the support of different parties on different issues, and fraction remarks from the committees is one way of signaling what is politically important, and what negotiable.

There has been a slight decrease in the number of laws discussed in the Storting and little change in the total time used on these issues (Olsen 1983, 47). But the legislative processes has changed in content, especially the last two decades, focusing more on frames (for example, framework laws) and less on details, leaving greater latitude for the administrative formulation of detailed regulations, a development quite like the one in the U.S. Congress. During this period, the time used on the budget has decreased, even though the Storting still uses relatively large amounts of time in discussing the budget in the different committees, and has changed in a more general direction, trying to change the attention of the representatives from details to principles (Arter 1984, 182–85). Since 1968 the number of budget recommendations has decreased from 200–300 to about 20. Since the 1970s the conflicts in budget issues have increased somewhat, reflecting the turbulent parliamentary situation, but still do not involve more than one-fifth of the issues.

Since 1945 the Storting has not spent much time in controlling the executive, in contrast to the United States (Olsen 1983). The tradition is to permit the government to organize its own internal apparatus. Besides the institutional and regular control of the executive, the Storting operates more according to a "fire alarm" principle and very seldom discusses more general control questions. The reports to the Storting from government about different public institutions such as executive agencies have become fewer and less specific over time, resulting in a decreasing workload, but probably also decreasing the insight necessary for effective control.

During the postwar period the number of questions posed to ministers has increased substantially. A weekly question hour was introduced in 1949, originally meant for the opposition (Arter 1984, 334–38; Olsen 1983, 49). The questions have primarily a specific and constituency-oriented character, especially for some of the backbenchers, since the general debates are more formalized and hierarchical. They are also important for the contact it provides with the voters, for representatives function in a "go-between" role. There were 85 questions posed during the mandate period of 1945 through 1949; 735 questions were posed during 1989 through 1993 (Rommetvedt 1994, 15).

The Storting has traditionally not had any public hearings, even though it has been discussed several times, stressing a norm of secrecy and an opportunity to work undisturbed by other actors, especially in committee. This was changed in 1996 with the introduction of such a system on a temporal and narrow basis. The Storting does not have any role in the formal approval of personnel, even though a reorganized committee for control and constitutional questions lately has attempted to control the procedures of the appointment of leadership positions in the public sector, albeit without much success.

Party Affiliation and Demography in Congress

The dominant party in Congress in this century, especially since the 1950s in both chambers, has been the Democrats. But Congress has a long tradition for coalitions between the parties, especially Democrats from the South with

Republicans. A trend is, however, that the ideologically motivated and party-unity voting in Congress has been growing over time, especially since the 1970s, stressing partisanship in the body. This change potentially counteracts the long trend of fragmentation and a lack of collective interest and perspective in the parties, although not in Congressional politics as a whole.

The length of careers in the Congress has increased 150 percent during the last 100 years, compared to a stability for the presidents and judges in the Supreme Court (Hibbing 1993, 67). But members of the last two congresses have experienced shorter careers and both politicians and voters are now advocating term limits. The conventional wisdom is that the members of Congress have a cyclical development in their careers, starting out with an aggressive, issue-oriented, close-to-voters style, finishing with a more internally oriented style characterized by less contact with and support from the voters (Hibbing 1993, 69). But it is empirically very difficult to discern this pattern. On the contrary, there seems to be small changes in the members' voting support, voting behavior, and contact with the constituencies during their careers. The only factor that seems to change is that the members' legislative effectiveness and influence increases with tenure.

Since World War II the Senate has experienced major changes both in the formal organization and the recruitment of senators. Overall, the Senate was a stronghold for the Democrats between 1958 and 1981, with the distribution of seats between the parties more even before and after (Ornstein et al. 1993, 14). The geographical pattern of voting for senators has also undergone changes. During the 1950s the Democrats were strong in the South and West, the Republicans in the East and Midwest. During the 1970s the Democrats had a majority in the Senate elections nearly everywhere except for the West, implying also a shift towards the direction of a more liberal Senate (Ornstein et al. 1993, 15). Since that time Republican strength has been increasing, especially in the South. Until the late 1970s, the Democrats senators gradually became more politically homogeneous and the Republicans more heterogeneous, a trend that reversed somewhat during the 1980s. But overall there has been increased partisanship in the Senate, reflected in stronger party discipline in senatorial voting behavior in the 1980s and 1990s (Ornstein et al. 1993, 17).

The turnover in the House of Representatives has been varying somewhat over time, but tending to decrease. During the 1970s it was rather high, and during most of the 1980s was low, a tendency quite the opposite compared to the Senate (Dodd and Oppenheimer 1993, 41). But as in the Senate, party conflicts in voting behavior have increased, meaning that the traditional coalition between Republican and Southern Democrats is of less importance. Additionally, the group of Democrats has itself grown more heterogeneous. Generally, the Republicans, as the minority party in the House for most the postwar period, held a strategy of cooperation with the Democrats. This trend changed however with the Republican presidents in the 1980s, creating a much more aggressive style, which was even more strongly emphasized when in opposition to the Clinton administration.

Party Affiliation and Demography in the Norwegian Storting

Since 1884 there have been many changes of political influence and coalitions of political parties in the Storting, but generally much political stability, a feature that seems to be related to the parliamentary principle and its pressure towards peaceful coexistence and collaboration between the legislature and the executive. There have been few periods of clear-cut dominance and more of some turbulence and minority government. The main exception from this is the "one-party-state" period during 1945 to 1961, which was dominated by the Labor Party (Seip 1963). The Labor Party has by far been the most important party during the postwar period, but its electoral fortunes have been mixed and weakened somewhat in the last two decades. Even though Norway experienced a swing to the political right during parts of the 1970s and 1980s, the nonsocialist parties had problems in establishing enduring coalitions. This trend is reflected now in a minority government, based on the Labor Party, a party that has never participated in any coalitions.

The political amateur has always played an important role in the Storting and the development towards professionalization has been slow but obvious, especially during the past several decades. This professionalization has been connected with more resources, higher education, and more politically specialized and relevant job experience (Olsen 1983, 51; Rommetvedt 1994). The amateur profile of the representatives has been related to the disproportional geographical representation in the Storting. Until 1952, the Constitution stated that two-thirds of the representatives must come from the periphery, and the over-representation of the periphery is still evident.

The occupational background of the members of the Storting has changed substantially since 1814. The first fifty years were dominated by civil servants, a group that experienced a sharp decrease in representation between the world wars, and a slight increase after 1945 (Arter 1984, 9; Olsen 1983, 52). Business leaders and the self-employed have traditionally been an important group in the Storting, but have decreased substantially during the postwar period. This is also the case for farmers and fishermen. Moreover, after 1945 the Labor Party has never had many manual workers among the representatives, but has been more characterized by representatives with experience from the public sector, the party, and the press, reflecting the typical new "professional politicians" (Arter 1984, 9). Teachers have experienced the largest increase in the period since 1945 and are now the largest occupational group in the Storting.

The Storting has never had aristocrats as members and never been a typical socioeconomical elite; it is more socially similar to its population than most other Western legislatures (Olsen 1983). The level of education in the Storting has increased substantially over time. During the two decades after World War II, 34 percent of the representatives had some kind of university education, a figure that had increased to 62 percent after the election of 1993 (Rommetvedt 1994, 7). The relative proportion of one traditionally strong group, the lawyers, has decreased sharply. The representatives have had, and continue to have, a solid background in the parties from different levels (Olsen 1983, 53). They also have a good deal of experience with interest groups and public offices at lower governmental levels, but at the county and national levels as well (member of public committees and alternates to the Storting, for example). Their local, geo-political background has also always been important for the members of the Storting, but is modified by party affiliation.

The parliaments in the Nordic countries have had for a very long time the highest percentage of women members of all the legislatures in the world. The Storting is no exception to this Scandinavian pattern. Its share of female representatives has doubled during the last 20 years and is now between 35–40 percent (Heidar 1983, 76).[4] This proportion also reflects the decision of many political parties that their election lists should have at least 40 percent women.

CONTEMPORARY INTERNAL STRUCTURE IN THE LEGISLATURES

The Principal Internal Structures in Congress

Today the Congress has a number of important features (Maidment and McGrew 1991; Wilson 1987). These reflect a long history of institutional development, as well as contemporary political pressures:

1. It is structurally more specialized—most importantly horizontally—than the executive and judicial powers.
2. Its bicameralism is originally based on different purposes and the two chambers have a different electoral basis. The Senate, indirectly elected and with longer election periods, was intended to balance the House of Representatives, where representatives are elected on shorter terms and are more voter-sensitive (Mezey 1991, 7). These roles remain important still today, even though both chambers have similar problems with capacity and constituency exposure.
3. It is a very decentralized, individualistic and open institution (Anagnoson 1989, 96; Thurber 1991, 5).
4. It has traditionally had relatively weak leadership, especially in the House of Representatives, compared to other, more party-based Western legislatures. During

the last decades the hierarchy appears more important in Congress in general, but also more divided among more leadership positions.

5. The importance of party affiliation and discipline is relatively low. The importance of voting along party cleavages is relatively lower compared to legislatures in other Western countries, but is increasing over time, especially among Republicans during the last decade (Rohde 1989, 138).

6. The backbone of its structure is the committee system, a reflection of the structure of the executive power, both the standing committees, and to an increasing degree, the subcommittees, where individual goals more actively can be followed (Fenno 1987, 160; Jacobson 1987, 74). Inside general frames, committees can choose rather freely their own internal structures and procedures (Fenno 1987, 149);

   The committee structure is more important and more detailed in the House, a pragmatic necessity for handling the workload, accommodating the number of representatives, and also is a reflection of heterogeneity (Mezey 1991, 24). Committees in the Senate have relatively less influence in the floor debates, generally develop less expertise, and have somewhat less-powerful leaders (Fiorina and Rohde 1989, 6);

   The committee structure has hierarchical features and is based on a modified principle of seniority concerning leadership positions. Generally, the leadership positions represent a powerful base in Congress, both related to internal lawmaking processes and contacts with actors in the environment. The members employ much of their available resources and attention in the committee, thereby shaping a more narrow profile of expertise and interests.

7. The floor debate and the deliberative aspect are generally not very important, but more so in the Senate than in the House. This is a reflection of a relatively low interest in the collective goals and purposes of the legislature.

8. The parochial orientation is obvious—that is, the constituencies and "home style" are crucial for the survival of members of Congress (Fenno 1978; Fiorina and Rohde 1989).

9. Congress actively uses public hearings for overview, control, and exposure of politically relevant information. These hearings are also important for individual members of Congress in impressing their constituents.

10. Congress has built up a strong resource basis, both collectively as a counter-expertise body, but also connected to the committees and the individual representatives (Jacobson 1987, 43; Polsby 1987, 102, 108; 1990, 17).

11. Generally Congress is supposed to be, and actually is, more active and taking more initiatives than legislatures in most countries, and does not just react to the actions of the executive (Mezey 1991, 23). The independence of Congress has increased in the 1990s as the "revolution" of 1994 has its effects on the relationships between the legislature and the executive.

The Internal Organization of the Storting

Some of the main characteristics of the contemporary Norwegian Storting are:

1. It is structurally far less specialized, especially horizontally, than the Congress. This is a reflection of the smaller size of the legislature, less-specialized executive power, and a election system based on one type of constituency, not two, as in Congress.

2. The members elected to the Storting are allocated to two chambers that have the same electoral basis, each chamber involved in different stages of the lawmaking process, but this way of organizing is of limited political significance (Arter 1984, 16, 310). This feature implies a more homogeneous legislature than in the United States.

3. It is obviously more centralized, collectively based, and characterized by consensus than the Congress, even though conflicts seem to have increased after the latest elections (Rommetvedt 1994).

4. The formal leadership in the Storting (i.e., the presidents) is weaker than in Congress;

5. The backbone of the Storting is the party groups and their leadership, more than the committee system—quite unlike the Congress—even though the legislatures have been more alike in this respect. The party groups coordinate the activities of its representatives in the committees and more generally, and the pressure for party discipline is very strong.

6. The committee system, a structure for the whole Storting, is a reflection of the structure of the executive power, but is moderately specialized, with no subcommittees (Arter 1984, 202). Generally, the committees have been of growing importance, just as in Congress, attracting a lot of attention from the representatives and developing expertise and "trained incapacity" to see the world in terms of the tasks of their own organization. A growing tendency of conflicts may, however, weaken them somewhat again, compared to the party caucuses (Rommetvedt 1994).

   The issues are allocated to the committees according to what type of sector or policy area they belong to. The committees are first of all working with the proposals coming from the government or the government bulletins (i.e., private proposals or bills mean almost nothing in the legislative process), quite the opposite to the system in Congress (Arter 1984, 97).

   As in Congress, the committee structure has hierarchical features and is based on a modified principle of seniority concerning the leadership positions (Hellevik 1969). Generally, the leadership positions in some committees represent a powerful basis in the Storting, but are more internally oriented than in Congress.

   A major difference between the Storting and the Congress is that party caucuses in Norway in many ways modify the importance of the committee system and the development of specialized individual careers. The party groups ensure that the activities in the committees are in accordance with the party programs and collective goals, and they move representatives between committees rather frequently, reflecting seniority and unequal importance of committees, but also avoiding too much specialization, parochialism, clientelism, and the development of individually oriented careers (Olsen 1983, 59–61).

7. The floor debate is generally not very important as a decision-making arena, but probably relatively more important than in Congress (Arter 1984, 156–57). The debates are very hierarchical and structured, produce few political surprises, but have more symbolic importance as some kind of "window-dressing," features that are also well known in Congress, but are even stronger in the Storting.
8. The parochial orientation of the representatives is of significance, but is of far less importance than the party and committee affiliation, and of obvious less importance than in Congress.
9. The Storting does not normally use public hearings, but mainly meets on an ad hoc basis, both formally and informally, with government politicians, civil servants, and different pressure groups for consultation and information, thereby appearing somewhat less open than Congress. This lack of exposure must be related to a more collectively oriented role for politicians, leading to less need of "selling" individual political attitudes and issues.
10. The resource base of the Storting has increased substantially during the last years, paid for both by the parties and the government, but it is more oriented towards technical needs than counter-expertise, reflecting the traditional "layman ideal" in Norwegian politics (Olsen 1983, 55–56). The resources are, however, generally on a much lower level than in Congress.

Summing Up the Similarities and Differences

The Norwegian Storting is generally based on many of the same functions as the Congress, namely, lawmaking, allocating resources, and controlling the executive. But there are also major differences, first of all related to a formal and informal, more-cooperative role with the executive, and that it is composed on quite another electoral basis—both effects of the parliamentary principle (Hernes 1971, 1977; Olsen 1983, 39–75). Structurally, this leads to obvious differences, like a much more specialized and complex structure in Congress.

One way of summing up the similarities and differences in the internal structure in the U.S. Congress and the Storting is to focus on the structuration of coordination and specialization in the bodies. Generally, in legislatures the parties and the more general committees are structural mechanisms of coordination, while specialized committees imply more specialized expertise, stronger orientation towards the particular constituents, and furthermore make it more easy for outside groups to find points of connection with government (McCubbins and Sullivan 1987, 83).

From our outline above it is quite obvious that Congress has put less weight on structures of coordination and reflects far more heterogeneity than the

Storting. The parties are of relatively less importance and the more specialized committees are numerous and probably outweigh the more general ones. In the Norwegian Storting the structural coordinating forces are much stronger, partly because the party groups are very important and partly because the committee system is far less specialized, and some general committees are very important (such as foreign affairs and finance).

Having now outlined the principal structural differences between the two legislatures, there is one especially important similarity between the two. The public agenda in both countries have expanded substantially, creating more work load and problems of capacity. This again is reflected in internal changes in the legislatures, including expansion of resources, greater structural specialization, more professionalization, and so on (Polsby 1987, 116). These features probably create greater activity, more conflicts, and an ambiguity of authority, but also encounter more efforts of coordination in the Storting than in Congress.

THE CONTROL OF THE EXECUTIVE BY THE LEGISLATURE

Congress Controlling the President

One important instrument of control of the executive branch by Congress has been the *legislative veto*. The background for this veto is that Congress in the process of delegation of authority to the executive branch needs an instrument that can control executive-branch organizations in their process of taking action, not only in the stage of initiation, but also after the actions (Sundquist 1992, 294–95). The legislative veto implies that Congress—both houses, one of them, or a committee—can review or reject a proposal from the executive branch before the actions are taking place.

Traditionally, the legislative veto was used in the reorganization of executive agencies, thereby obstructing the president from freely designing the federal bureaucracy. But twenty years ago Congress started to broaden the basis of the legislative veto, writing it into many laws, thereby making it more controversial and weakening the influence of the president. Given that this practice violates the separation of powers, the Supreme Court decided in 1983 (*INS* v. *Chadha*) that the legislative veto was unconstitutional.

This decision seems to have fostered two types of reactions from Congress (Sundquist 1992, 298–301): First, Congress decided upon a procedure of joint resolutions, demanding that proposals and plans from the president should be approved by both houses. Even though the issues were divided and the most important ones could go on the "fast-track," this new procedure probably obstructed the president's use of delegated authority more than the gradual spreading of the legislative veto before 1983. But this procedure also made the decision-making processes more cumbersome and complicated for Congress. Second, Congress has continued to use the legislative veto in certain issues after 1984, such as in the reprogramming of expenditures, because it believed that the executive branch would listen to its opinion in any case.

In the U.S. Constitution, the formal authority of the Congress concerning the declaration of war is quite clear, along with its abilities to make policy decisions and allocate resources to the military (Sundquist 1992, 303). But the development concerning the demand for taking rapid military action, in a complex and technologically advanced world, has in fact undermined this congressional authority and made the president a commander-in-chief more in his own right. The reaction of Congress to this, but at the same time a formal approval of the power of the president, was the War Powers Resolution of 1973. President Nixon vetoed the resolution, but it was overridden. The Resolution contains a procedure concerning the declaration and duration of war. The president must consult Congress before ordering the military into action and must report back immediately about the action taken, but apart from this he retains a good deal of discretion (Sundquist 1992, 306–7). However, a majority in both houses can stop the president from acting, and Congress has to give affirmative approval within sixty days; without this approval, the president is obliged to stop the military action.

Traditionally, Congress has a stronger constitutional position concerning the approval of treaties. Two-thirds of the members of the Senate must approve a treaty before it goes into effect, thereby limiting the president's foreign-policy discretion (Sundquist 1992, 310). During periods of bipartisan agreement in foreign policy this is no problem, but during the previous 20–30 years in the United States, foreign policy has been increasingly controversial. The rule requiring a two-third's majority makes it more difficult to create successful treaties, but on the other hand, since it actually requires bipartisan support, may be seen as a kind of democratic safeguard in important and far-reaching issues.

Another potential instrument of control of the president by Congress is the advise-and-consent procedure, involving Congress in the president's appointment of members of his cabinet, federal judges (including the Supreme Court), leaders in executive agencies, among others. This procedure appears to foster anticipated behavior and "sounding out" from the president, especially if candidates can create political embarrassment. Advise and consent is therefore especially an instrument for the opposition to oppose the president and his administrative program. The number of posts over which Congress avers its right of advise and consent has been increasing, so that there is ever more legislative involvement in the president's administration.

The budget process has always been a source of tension and conflict between Congress and the president. The Congressional Budget and Impoundment Control Act of 1974 was meant for Congress to counteract the increasing power of the president over the budget process. The budget reform established another structure of executive control: it created budget committees both in the Senate and House, and it created the Congress Budget Office (Gormley 1989, 157). This new structure was also related to new procedures for budgeting.

Thurber (1988, 101), in his assessment of the consequences of this reform, emphasizes that it strengthened the role of Congress in the budget process relative to that of the president, and forced the president to cooperate more with Congress. The reform also made the budget relationship between these two main actors more difficult and ambiguous and increased the number of participants in the budget process, especially through increased decentralization in Congress but also through the increased potential for coalitions between interest groups and Congress or the president. This change in budgeting also appears to have made the process more open.

The Gramm-Rudman-Hollings Act of 1985 was in a way a sequel to the budget reform in 1974, attempting to establish procedures for automatic cuts in expenditures to reduce the deficit (Thurber 1988, 103). Even though the Act originally was meant as a weapon for the Republicans to use against the Democrats, it eventually turned out to be an instrument for the Democrats in their control of a Republican president (West 1988, 96). Moreover, it provided Congress the opportunity to compete with the president in influencing the policy agenda. Like the legislative veto, parts of this legislation were declared an unconstitutional infringement upon the separation-of-powers doctrine (*Bowsher* v. *Synar*).

Members of Congress from either party can establish coalitions with the president through the coupling of issues to their own constituencies and thereby create a common identity (Sullivan 1987, 300). Generally, the more short-term conditions seem to have more importance for fostering coalitions than the more long-term ones such as party affiliation and position inside Congress. If the president is popular and the economic forecasts are good and it is early in the president's term, it seems to be easier to establish coalitions. But these factors seem also to point in the direction of dependence on the president—the downside of the potential for members to influence decision-making—gaining popularity in their constituencies, and increasing their chances of reelection. The stability in such coalitions appears to be lower if such members hold marginal seats and score low on trust among their constituents.

The Control of the Executive by the Storting

The parliamentary principle in Norway creates a closer relationship between the executive and legislative powers than the presidential system in the United States, because the executive is based on the majority in Parliament and is not elected separately. The probability of developing cooperation rather than conflict and tension is higher for many reasons: First, if the executive systematically undermines the authority of the Storting and engages in any wrongdoing in administration, it will lead to mistrust and nonconfidence in the Storting and change of cabinet, either during the election period or in subsequent elections.

Second, the parliamentary principle often makes the parliament a stronghold for the parties. The political center of the parties is there, and the executive must relate to the most important political actors, even if they are not directly represented in the cabinet.

Third, the executive will under such a system have people with solid parliamentary experience—often viewed as a major background for being a minister, a feature quite the opposite of the presidential system in the United States. In the U. S. cabinet, members are often strangers to the Washington establishment (Heclo 1977), although the development of policy communities has increased the connections of executives. Common political background for the main political actors in a legislature and the executive is suppose to smooth their relationship and create mutual understanding and fewer conflicts.

These are among the main reasons why the style in the legislative–executive relationship in Norway always has been, and remains, a relative cooperative one, characterized by trust and mutual adjustment, and why a broad system of sharing functions has not emerged. The general role definition is that the executive has to attend to decisions and signals from the Storting but has substantial discretion in doing so, and exercises executive oversight, but must report back to the Storting in a general way. The control from the Storting is more a general political one, rather than the specific ones through procedures in different policy arenas such as in the United States.

The formal connections between the Storting and government are as follows (Olsen 1983): The primary contact is through the party groups, especially for the party or parties in office. A separate power base for ministers is difficult to develop, since the activities in government and parties are coordinated and discussed closely.

Second, the political executive leaders participate in formal meetings in the standing committees in order to give information, answer questions, negotiate, and so on (Arter 1984, 319). This provides members of the Storting with the potential for mutual information and adjustment, but primarily for mutual understanding and close informal contacts.

Third, ministers participate in the floor debates in the Storting, where the proposals from the government are discussed, questions from the representatives are answered, and the ministerial responsibility exposed. Ministerial responsibility means that the minister is accountable to the Storting for everything that goes on inside his ministry and policy sector, even if he or she has not participated directly in certain decisions. If the Storting puts forward a motion for a vote of no confidence towards a minister or threatens to do so, the prime minister can either let the minister resign or declare that the whole cabinet will go collectively, if the opposition presses for no confidence (Arter 1984, 226, 230). This is far more politically serious and will often cause the opposition to be reluctant to press for the vote.

The Storting appears generally more preoccupied with the information side of the principle of ministerial responsibility than with detailed control of what is actually going on in a ministry. This is probably a reflection of problems of capacity and a variant of the "fire alarm" style of control. A minister attempting to misinform or avoid blame will always have more trouble than one informing the Storting about problems and taking the blame for failures. The informal networks between the legislature and executive are also important, supplementing the formal connections (Hernes and Nergaard 1990, 94–100). This system, labeled "power integration," is of special importance for both the executive and the opposition when the parliamentarian basis for the government is insecure—for example, during periods of minority coalition governments. Power integration can both secure a relatively stable power basis for the party or parties in power and provide flexibility, but can also provide influence to other parties, a feature that is mostly lacking if a majority government is in office for a long period of time.

In summary, the Parliament in Norway lives in peaceful coexistence with the government it has selected, as long as that government is following the Constitution and laws (Olsen 1983). This means that the Storting has a relatively limited number of instruments of control directed towards the government, and normally does not use these in any active way. The main mechanism of control is *ministerial responsibility* and that the cabinet has to quit if trust from Parliament is breaking down or withering away. Shared norms and self-control in the executive mean that a broader system of control from the Storting is not necessary.

The Control of the Executive Bureaucracy by Congress

Both the Congress and the president are trying to design organization structures that are aimed at controlling and reviewing the federal bureaucracy, resulting in a unique and complex administrative structure (McCubbins and Sullivan 1987, 403; Moe 1990). This is meant to channel and constrain the activities in the bureaucracy. This awareness of the importance of how the executive bureaucracy is structured is seen in many reorganization processes, where Congress and the president are among the actors in conflict (Gormley 1989; Hult 1987). The Congress is trying to structure the decision-making processes inside the bureaucracy through mandating particular procedures. This gives some potential for control and also for establishing cooperation between Congress and the bureaucracy in making public policy, a cooperation that can undermine the authority of the president over executive agencies. The committee system in Congress is especially important for establishing this "subgovernment" (Mezey 1991, 24).

The control of the bureaucracy from Congress is mainly connected to its lawmaking role. Its main means of control appear to be (McCubbins and Page 1987, 410–13): Defining the regulatory scope of the agencies, structuring the decision-making procedures (limiting alternatives, specifying due processes), creating incentives, having oversight over the activities of the agencies, and so on. There appear to be two types of oversight: "police patrol," characterized by members of Congress actively looking for problems, and "fire alarm," which involves Congress taking action when there are indications of a crisis (McCubbins and Schwartz 1987, 426). The latter type of oversight seems to be the more frequent style.

Attempts at increased control by Congress over the president and the federal bureaucracy seem to have many faces (Gormley 1989, 194). One increasingly important type is *legislative freelancing*, which involves individual members of Congress taking action towards the federal bureaucracy in a uncoordinated and fragmented way. Second, legislative oversight can be of the normally more institutionalized type (Kaiser 1988, 75). This category encompasses select committee investigations, specialized subcommittee hearings, the use of inspector generals and independent councils, casework, the intervention in bureaucratic decision-making processes on behalf of an individual, pork-barrel politics, interventions more oriented towards a district or state, entrepreneurial politics, the making of policies that many are benefiting from and few are paying for (for example, environmental and consumer policies, civil rights), and "sunset review," which is the automatic death of a bureaucratic unit or program unless the legislature is actively doing something before a certain time limit. A third type of oversight is *statutory specificity*—implying specific instructions from Congress such as priorities, standards, or time limits—that attempts to cope with vague and unsubstantive laws.

Sundquist (1992, 107) emphasized that the existence of many means of control by Congress has something to do with divided government:

> When Congress distrusts the executive branch, it is more likely to withhold discretion from administrators, write detailed prescriptions into law, impose constraints that may prove to be unworkable or result in the inequitable distribution of benefits, and intervene in day-to-day administration to the point of "meddling." It is also more likely to create quasi-administrative and supervisory bodies outside the executive branch (or even within the legislative branch itself) and informal relationships that make the legislature (or individual legislators) to some degree co-administrators of particulars programs, thus departing from the accepted notion—accepted, notably, by the Supreme Court and enforced by it when pertinent cases arise—that administrative responsibility should be fixed clearly and exclusively in the executive. . . .

The Control of the Central Administration by the Storting

The self-control in the central administration in Norway appears to be strong, reflected in strong norms of attention to political signals and small problems in dealing with the balancing of different considerations in the administrative role (Christensen 1991a; Jacobsen 1960). This may be one reason why other central actors seem to put so little weight on developing different instruments for controlling the central administration (Olsen 1983). Norway has, for example, no court expressly directed towards dealing with administrative questions.[5]

The Storting seems not to spend much time in controlling the central administration. It appoints a government board of auditors, in an economic-review position with an administrative apparatus connected, but this body seems in the past to have been preoccupied with economic details, even though it now attempts to broaden in a consequence-oriented direction and tries to objectively oversee the central administration.

The Storting has also created several ombudsman positions: for administrative questions, the military services, consumers, women's rights, and children. These positions must partly been seen as an instrument of control of the central administration, or as some kind of independent bodies' organizing systematic attention to underrepresented groups and interests (Egeberg 1997; Olsen 1983). One problem with these bodies, and they have much expanded during the last ten years, is that they have relatively weak formal authority. So one conclusion may be that they are neither very good instruments of control or defending weak interests, nor do they matter very much for changing public policies in desired directions. There seem to be a tendency that these institutions increasingly are involved in symbolic battles between some of the major parties.

Taken together, this means that the Storting generally has a much more passive role towards the public bureaucracy, both formally and informally, than does the U.S. Congress. First, the formal task of the cabinet is to organize its own administrative apparatus. Second, it is also acknowledged that the preparation and implementation of the public policies are delegated to the executive power. This means that the Storting takes a passive position in the daily overview of bureaucratic activities. The formal core in this relation is the ministerial responsibility, but this is not used as an active weapon of control. Third, the connection between the central bureaucracy and Storting is good, though more informally than formally defined, reflecting also that the bureaucracy has a high standing in the Storting. Members of the Storting frequently contact civil servants to get information on different issues, and there are formal meetings with the bureaucrats in the committee system. This pattern of contacts gives representatives a great deal of information and also develops shared norms and values. It is also worth stressing that Norway has a relatively open and easy-to-grasp administrative apparatus so that the public has a better chance of understanding what is happening in government.

THE RELATIONSHIP TO VOTERS AND CONSTITUENCIES

A traditional perspective on members of the U.S. Congress is that they, in their behavior and strategies, constantly are thinking about reelection (Mayhew 1987, 18; McCubbins and Sullivan 1987, 13). This perspective is represented in many studies of the relation between Congress and the president, federal bureaucracy, interest groups, states, or constituencies. It is believed that members' attitudes are individually oriented and vote-maximizing; features that are furthered by a relatively low party affiliation and a fragmented Congress. The representatives seem to have more influence over their own success in elections, and use an increasing amount of resources and attention to seek reelection (Jacobson 1987, 39–43, 73). The way Congress works and its capacity as a governing body are believed to be affected by how the representatives win and keep their political offices. One major problem is that Congress is responsive, but not responsible.

An increasingly central component of the lives of members of Congress, according to this reelection perspective, is their strong contacts with their constituencies. One aspect of this contact is said to be of growing importance, and that is that the representatives act in a kind of go-between role, connecting the constituencies and voters to the federal bureaucracy (Fiorina 1987, 30). This is a traditional role but has been more time consuming, as it has in many other countries, because the bureaucracy is expanding and is more complex for voters to relate to and influence. Members of Congress must raise money to seek reelection. One way to do this is to expose ones' attitudes in Congress by more openly showing the internal voting behavior, by open hearings in standing committees and subcommittees, and so on (Jacobson 1987, 74). This exposure again appeals both to constituencies and interest groups, but also has the potential for making members more dependent on these groups.

The members' sensitivity towards their constituencies appears to be of importance concerning their strategies and attitudes in the decision-making processes and relations with other actors such as the president (Sullivan 1987, 289). Trust within the constituency and a safe seat seems to foster more autonomy and a stronger position in negotiations. Further, the internal fragmentation of Congress, combined with vote-maximizing behavior, creates more conflicts, more problems with political control, a less independent legislature, complex sets of alliances and dependencies with the outside world, and results in a relatively low score on governance.

As emphasized, members of the Storting in Norway are far less preoccupied with reelection than members of Congress, since they in practice have safe seats for as long as they are interested. The election period is four years, the chances of not getting reelected are very small, and the individual politician is moderately exposed in the election campaign. Electoral campaigns in Norway are collectively defined, and it is the parties that compete in plural districts with proportional representation, not individual politicians. And candidates' expenditures are paid for by the parties, but they again receive financial support from the government. So generally there are relatively few instrumental reasons for the representatives, motivated by personal careers and the need for resources, to have a very close contact with voters and constituencies, either before or after election.

However, members of the Storting do maintain reasonably close contacts with the voters, thus reflecting a cultural feature of the system (Olsen 1983, 71). Their frequent travel home to their constituencies is paid for by the government. They cooperate with county and local authorities in their struggles in getting support, programs, and resources from the central government. They also represent and connect individual citizens to the central bureaucracy. Moreover, members of the Storting expose local problems through posing questions to the ministers in the formal questions hours in Parliament. This activity reflects an important tradition in Norwegian politics, namely, that geographically oriented politics is of substantial importance—a tradition verified by geographical electoral cleavages. But this is not a factor of importance for reelection on a personal basis. Also, Norway has always had a strong "layman ideal," the political amateur, stressing close connections between representative and voters, and maintains suspicion towards the professionalization of politics (Olsen 1983, 51). The close contacts also reflect the centralization in Norwegian politics, leading to local actors trying to influence the central allocation of local resources.

CONCLUSION

Our main conclusions concerning the legislatures in Norway and the United States are as follows:

1. The similarities of the legislatures include steadily increasing work loads, structural complexity and formalization, and growing staff resources trying to cope with increasing needs for capacity. These factors combined seem to foster even more activities in the legislatures and probably more conflicts, a characteristic more typical in Congress than in the Storting.
2. The U.S. Congress is, however, traditionally much more based on voters, constituencies, and individualism than the Norwegian collective and party-based tradition, even though the differences seemed to diminish somewhat during the last decades, with the growing importance of partisanship and party caucuses in Congress.
3. The U.S. Congress is less hierarchical than the Norwegian Storting, and the changes in leadership much more evident, going from dominance of committee chairmen to

more influence of both party caucus leaders and subcommittee leaders. This means also that the dual character (party- and committee-based) of Congress has been more evident during the last decades, something that has been typical for the Storting since 1884. But the Storting is characterized by caucuses controlling committee activities, leaving few opportunities for substantial individual committee-leader careers.

4. The U.S. Congress is also much more structurally complex than the Storting, and has become even more so over time, and is also developing a relatively much more heavily staffed organization. Congress is also more preoccupied than the Storting with internal structural reforms.

5. The Norwegian Storting has always been reluctant to establish formal contacts with the environment than Congress has, which reflects the more open checks-and-balances system found in the United States, and the more individual and constituency orientation of its members.

6. Looking back at the analysis of the political macrostructure of the two systems and the differences between presidential and parliamentary systems, this analysis of the internal structure of the legislatures adds a dimension to the discussion of governance. The fragmented character of the checks-and-balances system in the United States is connected to and furthers a fragmented and specialized internal structure in Congress, while the more homogeneous parliamentary system in Norway is coupled with a much more homogeneous Storting. In the United States, Congress and the presidency, even during periods of united government, are much more in conflict, developing constant differentiation and heterogeneous instruments of control and influence. The Storting and the cabinet are, relatively, much more to be regarded as players on the same team, with more weight placed on compromise and cooperation, thereby creating more homogeneity in the system, which is reflected in the internal structure of the Storting.

   Congress assigns much less weight to coordination—a paradox, since the needs for that activity are enormous, and is a reflection of relatively weak parties and party caucuses and an underdeveloped structure of coordinating committees or other bodies. In the Storting, with much less specialization and fragmentation, the party caucuses are very important for binding together the actions of representatives in the committees. The interest and tradition for collectivity are also observed in the status of coordinating committees.

   The difference between the two systems in fragmentation and forces of coordination has also been briefly discussed in pointing to differences in the relationship to the voters, a close and instrumental one versus a more distant and institutional one, but also in the different relationships to the executive. The following chapter shall describe and analyze the development and structure of the executive and thereby also deepen our understanding of the executive–legislative relationship.

NOTES

1. Much of the staffing of the ad hoc caucuses and other similar organizations was reduced after the 1994 Republican victory.

2. The last increase in representatives includes some members that are allocated on a national basis to make the representation of the different parties more proportional.

3. The data does not distinguish between remarks that are minor or remarks that stress principles.

4. During the last decade, many of the major political parties have had women as leaders.

5. In the United States, the Tax Court and Court of Customs and Patent Appeals are in essence administrative courts.

CHAPTER FOUR

The Political Executive

CHAPTERS 2 AND 3 HAVE outlined and discussed similarities and differences in the macropolitical structures of the United States and Norway, as well as the internal structure of the legislature in the two countries. Our conclusions have focused on some major differences in the homogeneity of the systems. The political macrostructure in United States is formally and actually much more divided, with more sharing of functions, a feature also related to the highly specialized structure of the legislature, creating greater internal fragmentation and conflicts, and assigning less weight to the large problem of coordination. In Norway the parliamentary principle creates a close and consensus-oriented relationship between the legislature and the executive, coupled with less heterogeneity of the internal structure of the Storting and more emphasis placed on structures and processes of internal coordination, especially the party caucuses.

This chapter will focus on the highest hierarchical levels of the executive, the central apparatus connected to the political leader of the executive, the president or prime minister. In particular, we will look at the cabinet and what we generally can label the staff of the political leaders. The central questions to be answered are: What are the main characteristics, structurally and demographically, of the cabinet and the staff of the political leaders in the two systems? What features characterize the development trends of the systems in these respects? Are the differences in the homogeneity in the two systems, already discussed for the legislature, deepened by the structure and functioning of the central parts of the executive, or are there factors that modify the heterogeneity, fragmentation, and conflicts in the American system and the homogeneity, coordination, and consensus-orientation in the Norwegian system? We begin by describing the development trends in the cabinet and staff of the political leaders in the two systems, and use this as a background for analysis of the contemporary structure and demography within the two systems. The chapter finishes with a short discussion of the control of the legislature by the executive.

Development Trends in the Executive

Developing the Presidential Staff in the United States

Originally the resources allocated to the president from Congress for the managing the executive establishment were rather limited, but they were not earmarked, and that permitted the president to use them both on cabinet secretaries and clerks (Hart 1987, 10). This style of appropriation reflected constitutional considerations in Congress, namely, that the president should have discretion in using his allocated resources. During the first half of the nineteenth century, Congress reacted rather slowly to ease the increasing administrative burdens of the president, but the presidents were seldom explicit about their problems. But during the second half of the century congressional funding for the presidential support staff increased gradually. In 1900, the president employed thirteen people in the White House administrative and clerical staff, and the staff started to change into a more political and professional direction, moving away from providing only technical support to become general advisors to the president (Hart 1987, 16).

A new increase in the presidential staff occurred around 1920, when, after the Budget and Finance Act, the president could propose his own staffing, and the number of employees reached thirty-one in 1922. Now Congress also formally allowed the president to borrow civil servants from the executive agencies on a temporary basis; until this time, this had been done in an informal way. This practice has persisted and that sometimes makes it difficult to know exactly how many people are working for the president. In 1929, Congress allowed another increase in the cabinet of the president, followed by an increase in specialization, leading to more politically important and visible cabinet secretaries. During this period some patterns, believed later to be typical for a large staff, appeared. Some secretaries attempted to shield the president from the environment and tried to monopolize his attention, and there were also internal conflicts in the presidential staff.

The activities of the federal government increased a great deal during the first years of the presidency of Franklin Roosevelt; he attempted to adapt the constitutional design from 1787 to the new and increased demands on the presidency (Berman 1987, 99). This led to the Brownlow Report, delivered in 1937, on the organization of the executive branch of government. The Report had many different proposals: Increasing the personal staff of the president, and increasing the staff generally by including budget, efficiency, personnel, and planning units within the executive branch; establishing a stronger merit system and reorganizing the civil service; reorganizing the executive branch and its 100 separate agencies under some few large departments; and reforming the fiscal system (Hart 1987, 24). The Report was seen as important both for making an effective presidency and as an attempt to increase the president's power.

Congress rejected many proposals coming from the Report, but accepted the reforms of the presidential staff. Therefore in 1939 the Executive Office of the President (EOP), including the new units and an expanded personal support staff of the president (White House Office [WHO]), was established. The EOP, motivated by overload problems, developed and strengthened the administrative means of the executive. But it also underlined the development of a presidential branch, partly separated from the more operational components of the executive branch—the executive departments and agencies.

The budget of the EOP was $2.8 million in 1939 and $121 million in 1986 (Berman 1987, 110). In 1955, there were 1,403 employees in EOP, compared to 2236 in 1972, not including the people on loan from the executive departments (Hart 1987, 41). The proportional increase in size in this period was three times larger than the increase in the executive departments. Inside the EOP, the increase between units has been uneven, since the Office of Management and Budget (OMB) (established in 1970 as a successor of the Bureau of Budget) and the White House Office together have 60 percent of the total number of employees.

Since 1939, forty new internal units in EOP have been established, but most of them have disappeared after a relative short period of time, following a change of presidents. Congress established one-third of the units, which were related to new policies and programs, underlining its priorities and its interfering in the organization of the presidency. The rest have been established by executive orders and by reorganizations acts and plans from the president. The influence of the president in reorganization of the EOP has been strong for a long time, but was weakened in the early 1980s by new procedures, caused by an unconstitutional ruling against some of the old ones. The EOP now has a broader managerial function than proposed in the Brownlow Report, leading to overlap problems both inside EOP and between EOP and the executive departments and agencies (Hart 1987, 48). Further, the budget control from OMB over the executive branch is sufficiently strong that additional conflicts may develop.

The White House Office has developed to be the most important unit in EOP, together with OMB, working as a managerial and oversight organization for the president, and having many units as satellites (Hart 1987, 96). WHO had thirty-seven employees in 1937, and increased from fifty-eight to 560 during the period 1944–1974, a 5-percent annual increase, mostly during the period 1955–1971; probably half of the staff added were professionals (Kernell 1989, 188). During the period 1986–1987, WHO had 450 employees in addition to people on loan from the executive departments (Berman 1987, 113). This expansion of the White House Office can be explained by increased pressure towards coordination and control of the executive departments and agencies, a role that the cabinet has not been able to fulfill. Also, there has been increased complexity in the pattern of contacts between the president and different other important actors, especially Congress, implying an important gatekeeping role for WHO (Hart 1987, 127).

The staff in the WHO has also become more differentiated over time, a reflection of increased responsibility, and more functions for the president. The increased responsibilities of the WHO include serving the immediate needs of the president, controlling and reviewing the bureaucracy, and duplicating the functions in parts of the federal bureaucracy (Berman 1987, 100). The increase in complexity is seen when comparing the administrations of Eisenhower and Reagan. The former had one level and eleven subunits in the WHO, while Reagan had four levels and some twenty-nine subunits (Kernell 1989, 191). The lower levels of WHO consist of line-units, specialists in external relations, while the upper levels are staffed with generalists who are engaged in policy planning, have responsibility for internal maintenance, and are monitoring the line-units.

Increased complexity in WHO has lead to subunits with different identities, increased internal competition, divided loyalties, the creation of personal power bases, and so on. The WHO is often seen as some sort of "counter-bureaucracy," because the members of the cabinet "go native" (Berman 1987, 108). The president must therefore develop some means of enforcing his policy priorities on the rest of the executive branch, and the various components of the EOP are central in that enforcement process. Presidential problems, such as the loss of control and corruption, have often been related to the expansion and differentiation of the WHO (Kernell 1989, 234). Congress has both wanted to develop an instrumental executive power, but on the other hand tried to control this development. In the last decades it has especially worried about powerful staffers hired in the WHO, who exercise their power outside the congressional confirmation power and influence.

Cronin (1975, 121–24) suggested different explanations for the expansion of the Executive Office of the President over time. He stresses the increasingly important international role of the United States, the increase in social problems, an abdication of Congress in critical decision-making (at least until 1994), a growing need for coordination because of more complexity and expansion in public policy, the mutual distrust between the president and the permanent bureaucracy, and the increased importance of interest groups. Concerning the development of WHO, presidents have consciously created its upper levels, moved many important units closer to the president and thereby created more conflicts of domain towards other important actors (Kernell 1989, 192–93).

During the 1970s, the growth of the presidential staff appeared to slow down, followed by cutbacks, a tendency evident under Reagan, and one experienced a larger growth in the executive departments/agencies and Congress. The EOP seems also to have changed somewhat in the last twenty years, towards a more partisan and nonprofessional role (Hart 1987, 128). This deinstitutionalized, more political role seems relatively to have strengthened WHO, as it also became more politicized, encompassing more policy-making staff. In the 1980s and early 1990s the WHO staff has been more hierarchically organized, thereby making the president more dependent on the chief of staff (Kernell 1989, 185).

Changes in the Organization of the Prime Minister's Office in Norway

The Norwegian Constitution of 1814 does not say much about the prime minister, since the political leader was the king, and formally he or she has no hierarchical authority over the other members of the cabinet. Historically the position of the Norwegian prime minister has, however, been relatively strong after 1884. This strength is based in the parties and their programs, and defined by the role as a central political organizer (Olsen 1983, 81–82). The prime minister has always had close connections with other central political actors, like the parties, the party groups in the Storting, and the leaders of large interest groups. All prime ministers have also been party leaders, thereby giving them a solid political platform.

For a long time the prime minister had no regular staff, and would head another small ministry from time to time before 1940 (Berggrav 1985, 32). During the 1950s a small office for the prime minister was established (Arter 1984, 122). Since that time the staff has grown slowly, mostly in the last decade. The prime minister's office employed nineteen people in 1979, and the number now is between thirty-five and forty, of which six are politically appointed and about ten are in significant, administrative leadership positions. It has in the last decade been more specialized, both the political advisers and the professional staff, but also has placed more weight on internal hierarchical leadership. The first responsibility of the staff is helping the prime minister with information, contacts, and coordination, and the political influence of the staff seems generally to have increased. Given this trend, the staff appears to be able to increase their influence through strong prime ministers such as Gro Harlem Brundtland was before she stepped down in 1996.

The political leadership has been reluctant to develop the prime minister's office into a superministry, even though the needs and ambitions of coordination have grown somewhat. One example of the ambition was its active role in the adoption to the European Union, through participating in coordinating and influencing the negotiation processes (Christensen 1997b). The tradition under the prime ministers from the Labor Party has been that the prime minister actually selects the cabinet, even though the king has the formal authority to do that. In so doing, the prime minister consults the party, the party group in the Storting, and the labor union leaders. In selecting the team, the prime minister must take into consideration the experience available from the Storting (not too many though, because that will weaken the party group), geographical background, sectoral background, gender, and other relevant political characteristics.

Development Trends in the U.S. cabinet

The cabinet in United States has existed since 1789, even though there was no mention of the institution contained in the Constitution (Berman 1987, 100; Cohen 1988, 8). It was not controversial to have a cabinet, copying the experience in England, since this was seen as a natural way for the president to relate to the leaders of the executive departments. But it was not quite clear what tasks the cabinet should perform in a presidential system, especially in its relationship to the president and Congress, and the interaction of the two branches.

Even though the cabinet from the beginning lacked a firm constitutional basis, its position was not altogether ambiguous. Its role was constrained by the formal separation of power doctrine, but dependent on the actual development in the relationship between Congress and the president. The legislature had from the beginning, as stated in the Constitution, some functions related to the cabinet. Appointment of members of the cabinet must be confirmed by the Senate, and Congress has a central function in impeachment, while the removal power of Congress other than through impeachment has been more ambiguous and controversial (Berman 1987, 107). The Constitution also stated that no one could be a member of both the cabinet and Congress, thereby preventing the president from creating a cabinet composed of members of Congress, and thereby creating a quasi-parliamentary system. This feature has tended, however, to weaken the political expertise and experience of the cabinet, probably a disadvantage for the president. The cabinet was from the beginning torn by different considerations: Should it emphasize the capacity to act or to be a representative body, even though it was an executive apparatus (Cohen 1988, 4–17)? Should the members of the cabinet emphasize fulfilling the collective goals of the president and control the bureaucracy, or represent special interests in their departments or related interest groups, and attempt to influence the president on their behalf?

The establishment of executive departments, and therefore increases in the size of the cabinet, has occurred in waves. The first wave consisted of six departments—State, War, Treasury, Attorney General, Postmaster General, and Navy (Cohen 1988, 19). This first wave encompassed most of the departments traditionally defined as the "inner cabinet departments," close to the president and representing general values and norms and the "defining activities" of government (Cronin 1975, 76–86; Rose 1976). The second wave, known as the "interest group-wave," related to different economic sectors, came in the second half of the nineteenth century. It consisted of the Departments of Interior, Agriculture, Commerce, and Labor. These departments have later been called the "older, outer cabinet departments." The last wave started in the 1950, representing the interests of broad classes of citizens and related to conflicting interest groups, and concern the relatively small but important welfare state in the country. They have been called the "newer, outer cabinet departments." The two last waves have established departments that represent special interests, both providing potential instruments for the president to utilize and through which to build coalitions, but also potentially undermining the authority of the president by favoring special interests and creating fragmentation in the cabinet.

Some of the principal characteristics of the U.S. cabinet were established early in its history. It was from the beginning heavily dependent upon the president, represented no clear collective force or team, encompassed assistants or advisers with no independent power basis, and had no subleadership (Berman 1987, 105; Cohen 1988, 26; Maidment and McGrew 1991, 77). But the president has always had to take into consideration both the confirmation power of Congress and the necessity of representing regions/states and interest groups when composing his cabinet. The creation of "balance" within the cabinet is especially typical of the Clinton administration and other Democrat presidential administrations. The cabinet is not itself a complex political body, but has gradually increased in complexity. This is partly related to the increase in number of departments, meaning a growing specialization, and a development with more special units such as the cabinet councils (Cohen 1988, 28). The cabinet as an entity also traditionally has had few administrative resources.

Traditionally, the cabinet has had no common identity, no firm institutional development, but more identification with special interests, resulting in fragmentation and conflicts with the president, and between members of the cabinet. One trend in the opposite direction is that there seems to be, under some presidents, a growing number of cabinet members not characterized by being interest-group representatives. Rather, they are recruited to be experts and managers loyal to the president, and stressing more control and collectivity through the cabinet (Cohen 1988, 33–34). The cabinet has lacked a collective authority and lost power and functions to the Executive Office of the President. But some presidents have revitalized the cabinet, on a short-time basis, to manage policy-coordination and budgetary cutbacks.

Cohen (1988, 44), in analyzing the social basis of the cabinet from 1789–1984, categorized the development into three eras, each having different needs that were reflected in its composition. The first era is the founding and creation era, ending with the Civil War and Reconstruction. The second, the reform–progressive era, begins towards the end of the nineteenth century and ends around 1930. And the third is the big-government era, started by Franklin D. Roosevelt and continued during the administration of Lyndon Johnson. The nature of subsequent cabinets has been too varied to categorize easily.

The cabinet has always been a social elite, but, given this characteristic, has changed in many ways. The founders were younger than the cabinet members coming later and had a much longer tenure. During the last 130 years the average age of the members has been stable, somewhat over fifty, and the members have had tenures of two to three years—not very long careers! Most of the cabinet members have attended prestigious schools, mostly on the East Coast, with a marked increase in people from graduate schools over time, reflecting a long-time trend in education. There are small differences between the parties in the educational level of cabinet members, with Republican members tending to come from more prestigious schools.

On average, the cabinet members have an occupational background characterized by government jobs more than private law and business jobs, but the importance of this experience have changed a great deal. Government jobs as background has diminished dramatically, while business background, typical for the Republicans, has increased substantially, especially after World War II (Cohen 1988, 74). The same trend is evident for the last jobs held before becoming cabinet members, but here government background is still dominant. The occupational background varies greatly between departments: State Department and Treasury are characterized by government background, law is also strong in State Department and in Defense and Justice, while a business background is strongest in Treasury, Defense, and Commerce (Cohen 1988, 127). The typical interest-group departments are Commerce, Agriculture, Labor, Interior, and Education. Of the people having backgrounds in government, few have been members of Congress; many, but a decreasing share, have had state or local government jobs, and an increasing number come from the federal bureaucracy.

cabinet members generally have had a lot of different political experience, occurring in a cumulative pattern, dominated by having held public offices or been candidates for public offices. People with occupational background in government are scoring highest concerning former political activities, while representatives coming from business are scoring the lowest, also reflecting a difference in favor of the Democrats. Since World War II, the cabinet members' background in activities related to political parties seems to decrease, as a reflection of the more modest role of the parties, while the affiliation to interest groups seems to increase in importance (Cohen 1988, 119).

There is not much data comparing the cabinet members in the United States to other countries. One study by Blondel (1985) indicates that U.S. cabinet members are characterized by nearly the lowest average tenure in office among the Western countries and by scoring highest on the percentage of ministers having only one post in their ministerial career (Blondel 1985, 86–88, 104, 218). This is supposed to reflect the macrostructure of the system, with two competing parties, the term limits in office for the president, and the loose coupling of the ministerial careers to the Congress.

The Development of the Norwegian cabinet

The Norwegian Constitution of 1814 gave the king executive leadership, the right to choose advisers—the cabinet—and organize the bureaucratic apparatus. The king had some few Norwegian ministers together with him in Stockholm, called "the ministerial division in Stockholm," while the ministries and the main group of ministers were in Christiania (later renamed Oslo), called "the Norwegian government" (Debes 1950, 11–12). The Stockholm division consisted of a Norwegian prime minister and two other ministers, the latter rotating each year between this division and the one in Oslo. The leader of the Norwegian government was either the vice-king, a "statholder" (normally one from the aristocracy), or the oldest of the ministers, the latter one called *prime minister* since 1873. This leadership structure created a lot of conflict between the countries.

The king, representing the dominant country in the union, had a major influence in this period. The members of the cabinet were in something of a cross-pressure situation. One the one hand, they should be loyal to the king and prepare and implement his policies. On the other hand, they were career civil servants and not politicians in any real sense, having their loyalty to Norway and their connections to their own social and professional groups—the civil servants that dominated the central administration and the Storting. This meant that the main cleavage was that between the king and the Storting, not between the cabinet as such and the Storting. This cleavage grew wider after the 1850s, when the peasants mobilized and the Storting took on a more politically active role.

During the period 1814–1884, there were all together sixty-seven ministers and prime ministers in Norway. Thirty-five of these ministers were educated in law, demonstrating the dominance of the civil servants, and another twenty-two came from the military services (Debes 1950, 17). Thirty-two of the ministers had parliamentary experience, either from the Eidsvold assembly that created the Constitution in 1814 or from the Storting. This was, as Seip (1963) described it, a "civil-service state."

The breakthrough of the parliamentary principle in 1884 gradually changed the cabinet into an executive committee or agent for the Storting, and the king to a ceremonial leader (Olsen 1983, 77). Since 1884 the executive leadership has been based on party support, and it has been of importance whether there has been periods with majority or minority parties, single parties or coalitions in power, disciplined parties or not, and so on. This partisan basis of the cabinet in Norway is one important difference between it and the U.S. cabinet and presidency. It has not been unheard of for a president to include individuals with little or no partisan affiliation in a cabinet, but this would be almost impossible in Norway.

The role of the cabinet in Norway has since 1884 been a central one politically. It is a political arena in which problems are detected, worked with, and sometimes solved (Olsen 1983, 102). It is also central for solving political conflicts or deciding how to avoid conflicts. The cabinet is a "clearinghouse" for the coupling of problems and solutions, for "sounding out" or shaping a political profile.

Norwegian cabinets have always struggled with its dual role, in a way more than in the United States, since the collective aspect of the cabinet means much more: On the one hand, cabinets in Norway have traditionally functioned as teams focusing on *collective goals* with norms of consensus, meeting relatively for long periods together, placing more weight on work than on the routines of politics, like the budget process and important laws (Berggrav 1985, 39–42). But, on the other hand, the specializing forces in the cabinets have grown relatively stronger, especially after World War II, with increases in resources and specialized expertise. The ministers must use much more time than before in negotiations with strong special interests and function more on a contextual basis, according to a "fire-alarm" logic (Olsen 1983).

The cabinet structure from 1814 was a function of the type of ministries and ministers required to meet the basic needs of a rather passive state. During 1814–1815, there were six ministries: one for church and education, one for judiciary questions (five years later it merged with the police), one for police, one interior, one for finance and commerce, and one for military services (divided in 1846 into one each for the Army and Navy, but later remerged) (Debes 1950, 27). Relatively few changes of this structure occurred during the next one hundred years, except for changes during the world wars. The Ministry of Public Works (1885), Ministry of Agriculture (1900), Ministry of Foreign Affairs (1905, after dissolving the union), and Ministry of Social Affairs (1913) were added to the cabinet.

The horizontal specialization in the cabinet has increased greatly, with the establishment of many ministries after World War II. This specialization resulted partly from building ministries around new policy areas, either to further growth or to cope with problems of growth, partly by differentiation in already existing ministries to cope with problems of capacity: ministries for fisheries, industry, and communications during the late 1940s, ministries for the wages–prizes and families–consumers in the 1950s (later divided into families–children and administration), ministries for environment and energy during the 1970s, and ministries for culture and underdeveloped countries during the 1980s.

The Norwegian cabinet has never been very large, and there has been a tendency in the last decade to rationalize the structure into fewer ministries, sometimes with the old ministries becoming separate units inside one superministry. An incoming Labor cabinet in 1996 is, however, even more specialized.

The Norwegian cabinet has since 1884 been characterized by a collective style (Olsen 1983). Different members of the cabinet, especially those from the important ministries responsible for coordinating activities, participate normally quite a lot in the discussions and have a lot of influence. This is true even though there is also a norm of giving the minister with special knowledge more influence than others. This collective feature has been increasingly evident, especially after World War II. The members of the cabinet use more time together than most cabinets in the Western countries. They gather in a "cabinet conference" twice a week, the second one of these started by a preparatory meeting to the formal cabinet meeting the next day, headed by the king.

The Norwegian cabinet has not had a tradition of being divided into subcommittees, quite unlike most other Western countries. This seems to be a reflection of the collective and egalitarian style in the political system. The first cabinet committee was established in 1946 and discussed the first national budget after the war. The number of such committees has, however, remained rather small since then, although somewhat higher in coalition cabinets where the need for coordination is higher (Berggrav 1994, 65–70). The activities and influence of these committees have also been relatively minor, except for some few special periods, like the Europe subcommittee during the negotiations with the European Union.

There are two types of cabinet committees: the *pure* ones and the *mixed* ones. The latter indicate, for example, that some cabinet members are appointed as members in committees with representatives from interest groups. In the current cabinet there are five committees with members only from the cabinet: national security, science, Europe, health and social, and industry and energy. The cabinet committees are supplemented by committees consisting of undersecretaries of state, a feature reflecting problems of capacity for the ministers and the need of coordination. Today, there are six such committees: wages and employment, environment, Northern Norway, economic criminality, long-term policy planning, and information technology.

Traditionally, the cabinets in Norway have been a mixture of generalist and specialists, and of veterans and "guests." The members of the cabinet have had strong ties to the Storting, but party-tactical and substantive political reasons have also always led to a substantial proportion of members without such experience. Between 1814 to 1884, 44 percent of the ministers had no experience in the Storting, while the period from 1884 to 1945 was characterized by more ministers having legislative careers, nearly half of them six years or longer (Laegreid and Roness 1983, 23). The postwar period has the relatively highest score for the ministers without experience in the Storting, with about 40 percent on average having such a background (Heidar 1983, 66; Olsen 1983, 66, 93).

Norwegian ministers have traditionally had a very strong local political background. But we can differentiate this tendency across time: The share of ministers having experience as mayors had sharply decreased after 1945, while the proportion of ministers having been members of local councils has increased (Heidar 1983, 66; Laegreid and Roness 1983, 23). This change is probably due to more differentiated political careers of local politicians (Eliassen 1985, 124–28). Strong experience with the party on different levels of government has also been typical for ministers.

Socially, the ministers have been more biased relative to the population than the members of the Storting. Ministers from Oslo and the surrounding region have been a dominant group. This is true both concerning their place of birth and their residence when recruited, even though this pattern is decreasing slightly over time (Heidar 1983, 63). The relative share of ministers with academic backgrounds also is decreasing slightly over time (Heidar 1983, 64–65). This change reflects the decrease in a former dominant group in the cabinets—the higher civil servants. Their relative percentage decreased from 56 percent during the decades after 1884 to approximately 20 percent today. Lower public employees have, however, increased from 13 percent to approximately 25 percent during the same period. Employees from interest groups and political parties also have been increasing their share of the cabinet.

Traditionally, law has been the dominant educational background of ministers. During the period from 1945 to 1980, jurists comprised 38 percent of the cabinet (somewhat higher for Labor Party cabinets), a decreasing percentage over time (Laegreid and Roness 1983, 33). An economics background for ministers increased after 1945, but philologists and agronomists have also comprised important educational groups in the cabinets. There has been a differentiation of educational background according to type of ministry. After 1945, ministers educated in economics dominated or had strong positions in the ministries of commerce and finance; jurists held similar positions in industry, interior, justice, and foreign affairs; agronomists in agriculture; and philologists in education and church (Laegreid and Roness 1983, 21).

Before 1935, no women had been a member of the Norwegian cabinet, while the figures were 6 percent and 16 percent during the period between 1935 to 1965 and 1965 to 1980, respectively (Heidar 1983, 63). During the last decades this percentage has increased to around 40 percent, one of the highest in the world. Further, Norway had a female prime minister—Gro Harlem Brundtland—who was heading three cabinets and had a total tenure in this position of over ten years before she left office in 1996.

The average tenure for ministers during the period from 1884 through 1920 was two-and-a-half years, which decreased to two years during the more turbulent period of 1920 through 1940 (Heidar 1983, 16). Since 1945, 40 percent of the ministers have served in more than one cabinet and the average tenure has been about three-and-a-half to four years (decreasing over time, and higher for members of Labor Party cabinets), while 45 percent have been ministers for more than five years (Laegreid and Roness 1983, 20; Olsen 1983, 91). These are much longer ministerial careers than in the United States. Historically, about three-quarters of all ministers have served only in one ministry, primarily in the typical specialized ministries, while ministers with long careers often have been generalists and also are often the most central actors politically.

A large study of ministers in the political systems in Western Europe during the period from 1945 through 1984 can fix the profile of Norwegian ministers in perspective (Blondel and Thiebault, 1991). Relatively fewer Norwegian ministers have had experience in the Storting during this period, compared to the average among Western European countries, while the members with experience appear to have a profile comparable to the average (de Winter 1991, 48). This feature seems to reflect the tradition of balancing the cabinet; that is, it is important to have ministers both with experience from outside the Storting, with different geographical and/or a special competence/interest, but also not to weaken the parliamentary groups too much, demonstrating the political importance of the Storting.

The special significance of the coupling in Norway between the numerical and corporate channels, the integrated participation, is also seen by the fact that Norway by far is scoring highest among the Western European countries in the affiliation of ministers to interest groups (both having been national officials in general or in trade unions and employers' organizations more specifically) (Cotta 1991, 187). Sixty-four percent of the ministers during this period had been or were national political leaders, compared to an average of 43 percent, showing quite clearly that the cabinet in Norway has a firm party basis (de Winter 1991, 48). Norwegian ministers score much higher than their Western European counterparts in the average experience from local and regional politics—74 and 52 percents, respectively. This is especially due to their experience obtained in local assemblies (Thiebault 1991b, 33, 36).

Norway also shows some different tendencies concerning the comparative social profile of cabinet ministers. During the period between 1945 and 1984, 66 percent of the Norwegian ministers were university educated, compared to the average of 77 percent, a reflection of the dominance of the Labor Party for much of this period (Thiebault 1991a, 26). Norwegian ministers also have more varied backgrounds concerning their types of education as compared to most other Western European countries, meaning also that the percentage of jurists is far below the average. Norway scores much higher than average concerning woman in its cabinets, along with the other Nordic countries (Thiebault 1991a, 25). Norwegian ministers scored slightly below average on tenure during the period between 1945 and 1984, and they seem more likely to have held only one ministerial post, compared to the average (Bakema 1991, 75–90).

THE CONTEMPORARY STRUCTURE OF THE EXECUTIVE POWER: A COMPARISON

Main Characteristics

The executive power in political systems very seldom has the authority and power to dominate other important political actors. Normally, the executive power is only one, but very important, actor in a network of political and administrative actors. This network constrains the decisions and actions of the executive power, thereby putting stress on both its controlling power and cognitive capacity when designing an institutional system aiming at serving the goals of the executive (Moe 1985, 241–44). Our point of departure here, shown in the historical overview, is that the network of important actors is more complex and problematic for the political executive power in the United States than in Norway.

The modern version of the presidency of the United States is the result of an ambivalence towards the organizing and execution of political power. On the one hand, it contains both a formally and actually active and strong president concerning the main functions of executive power, receiving political demands, initiating policies and setting the political agenda, and taking an active part in lawmaking (Maidment and McGrew 1991, 70). On the other hand, the president is formally and actually constrained in different ways by Congress and the judiciary power. Taken together, the general plurality of the system and its different constraints make it difficult to design the executive structure according to a holistic or collective perspective (Moe 1989, 279).

An important feature of the presidency is its internal structural heterogeneity, both between the political leadership and the executive departments and agencies, but also inside each of them. One reason for this specialization and heterogeneity is, of course, the increased workload and increasingly complex political issues, resulting in a substantial growth in resources both in the Executive Office of the President and the federal bureaucracy. Another reason is the checks-and-balances system. The president must relate actively with Congress, and political and administrative capacity and more specialization are central ways to strengthen the political power of the president vis-à-vis the other institutions.

Another effect of the checks-and-balances system is that the executive agencies participate in "iron triangles" with Congress and interest groups. These foster and reinforce clientelism and thereby create a major problem of control for the president. This is in addition to the general problem of political control over a vast and specialized federal bureaucracy. One central answer to these problems of bureaucratic control from the presidency has been to increase the numbers of political appointees, both inside the Executive Office of the President, subcabinet appointees, and in leadership positions in the independent agencies (responsible directly to the president), and to develop numerous formal mechanisms for control (Maidment and McGrew 1991, 75).

In Norway, the executive power is much more collectively defined than in the United States, and it is much more common to talk about the government or cabinet than about the prime minister when focusing on the political leadership. These collective values are also reflected in the fact that the formal authority of the prime minister is weaker than in many other parliamentary system. But the parliamentary practice has generally developed a relatively strong political leadership in Norway, though varying somewhat with the actual partisan constellation. Concerning the main tasks, the political leadership in Norway has a similar profile to that of the president in the United States but the system is more centralized, thereby weakening somewhat the different demands for using resources in negotiations.

Generally, the central executive power in Norway is much more *structurally homogeneous* than in the United States, both in the political and administrative meanings of the word. This is, of course, a reflection of differences in size and cleavage structure, but also of the existence of "the Parliamentary Chain" (Olsen 1983). First, it is generally accepted that the Storting is relatively modestly involved in the preparation and implementation of policy. This limited policy role is reflected in a relatively small administrative apparatus and expertise in the Storting (Rommetvedt 1994). This means that the problems of capacity are addressed primarily through an expanding, although not vast, central bureaucracy.

Second, there is formally a strong hierarchy in the executive, between the political leaders and bureaucracy, and inside the bureaucracy. Moreover, the bureaucrats have built-in political considerations in their roles (Christensen 1991a). This means that the need for strong, daily political control of the bureaucracy is much lower than in the United States. Therefore, there are also many obvious structural differences in the organization of the political leadership: The capacity of the prime minister's office is modest, both politically and administratively. The number of political appointees in the ministries is very moderate (only sixty to seventy altogether), and there are no political appointees in the leadership of the executive agencies—the directorates; most of the civil servants are in the central bureaucracy.

The Power of the Head of Government

King (1993) discusses different variables for comparing the positional or organizational power of the head of government in political-administrative systems. He focuses on different sources of power, and we will use most of them to compare and sum up the similarities and differences between the United States and Norway.

1. *The Constitution:* Is the formal power of the head of government stated very clearly in the Constitution, and is this of importance for the political practice? Does the head of government have special means of power—means that are strengthening his or her formal position?

   United States and Norway are rather different on this variable. The formal, constitutional power of the president of the United States is strong, both concerning the formal authority in the separation-of-powers system and with regard to the legislature and the judiciary. The constitutional practice, especially in this century, has strengthened this position. In Norway, the formal, constitutional power of the prime minister is rather weak, barely mentioned in the Constitution. The constitutional practice, the parliamentary principle and its development, has created a relatively strong prime minister, leaning on the party basis.

2. *Direct Elections:* Is the head of government elected through direct elections or not? It is assumed that a political leader elected directly has a stronger basis of legitimacy, but probably also is more vulnerable towards changes and turbulence, than is a leader of a party or a leader selected as a result of a parliamentary process.

   Again, Norway and the United States are quite different. The president of the United States is elected directly, or more correctly, through an electoral college, while the prime minister in Norway is selected as a indirect result of an election or changes in the Storting during an election period. This means that the president bases his leadership much more on the direct support of the voters.

3. *Electoral Security in Office:* Are there any limitations of the number of electoral terms in office for the head of government? The thought is that the longer potentially in office, the more solid the power basis.

   The situation in the United States is that the president can be reelected for a second term, giving him potentially eight years in office. In Norway, as in many parliamentary systems, there are no formal rules about reelection, but the actual changes of the prime minister are more frequent. But there is, of course, the possibility for such a system to have a dominant party in certain periods, like the Labor Party, that is creating a stable platform for the head of government both in and between many election periods. Prime Minister Gerhardsen, for example, was in office for most part of the time between 1945 to 1965, and Prime Minister Brundtland had a total tenure of over 10 years, between 1981 and 1996.

4. *Effective Leader of a Political Party:* Is the head of government also the leader of a party? Further, is he or she the leader of a large and homogeneous party?

   In the United States, with weak political parties, the president is somewhat decoupled from the party as a power basis and therefore has a more narrow power basis than in Norway, where the importance for the prime minister of the party as a basis is very strong. Very often, the leader of the parties is also the parliamentary leader, and normally, after elections, the leader of the largest party or leader of the largest party in a coalition is assigned the task of forming a cabinet.

5. *The Control Over and Influence in the Legislature:* Has the head of government relatively strong influence in the legislature or not?

   In the United States, through the separation-of-powers system, the formal position of the president towards Congress is relatively weak, even though he has the veto power and other means of influence. In Norway, the potential power of the head of government, through the parliamentary principle, is higher, and even more so since the prime minister is also often the leader of the party, or very influential in the party. If the parliamentary situation, however, is turbulent and insecure, this potential is more problematic to translate into action.

6. *The Control of Appointments:* Does the head of government control the important political and administrative appointments or not?

   In the United States, the appointment power of the president is rather strong, but not without limitations through "advice and consent," resulting in conflicts from time to time with Congress, postponement of appointments, or direct rejections of the candidates proposed by the president. In Norway, the power of the prime minister and cabinet to appoint political, judicial, and administrative leaders is nearly unlimited. But it is important to emphasize that the judiciary plays no political role, the number of political appointees is relatively lower, and the heads of the executive agencies—the directorates—are not politically appointed. This means that the potential of political influence through appointments is relatively lower in Norway than in the United States.

7. *Controlling the Structure of the Executive Branch:* Has the head of government the power to organize and reorganize the executive branch?

   In the United States, the control by the president over the structuring of the executive branch is formally substantial, but also is divided between the president and Congress. It is, however, difficult for the president to cope with the complexity of the governing system, and he can only directly involve himself in relatively few changes. In Norway, it is constitutional practice that the prime minister and cabinet organize their own apparatus. If a restructuring is coupled with changes of laws or more money, the Storting must participate in the process but seldom does so very actively, except for a few highly politicized cases (Christensen 1997a).

8. *Controlling the Central Bureaucracy:* Has the head of government control over the decision-making processes and activities in the executive administrative apparatus (i.e., how autonomous are the ministries/departments and agencies)?

   In the United States, it is believed that the control by the president of the federal bureaucracy is rather slight, as a function of the structural complexity, internal conflicts and tensions in the executive branch, the existence of "iron triangles," and the powers of Congress and interest groups. In Norway, this type of control is potentially rather tight, since the hierarchical element is evident. It is formally and in reality the job of the prime minister and the ministers to coordinate and control the central administrative apparatus. But on the other hand, the actual autonomy, both in organizational and individual terms, of the bureaucracy is rather high. One principal reason for this is that civil servants know very well how to balance political and professional considerations (Christensen 1991a).

9. *The Size of the Staff of the Head of Government:* Is the political and administrative apparatus, closely connected to the head of government, large or not?

   In the United States, the Executive Office of the President, as discussed below, is huge, while the prime minister's office in Norway is rather small. This seems both to create the potential for greater control in the United States, but also more problems of duplication and overlap.

10. *The Existence of Collegiality in the Political Leadership:* Is the head of government formally and in practice the obvious hierarchical leader, or is the political leadership characterized by collegiality?

    In the United States, there is a very strong hierarchical element inside the political leadership: the president is relatively much more powerful than the prime minister in Norway, where the cabinet is really an important collegial body for coordination and control, both helping and constraining the prime minister.

The Executive Office of the President and the Prime Minister's Office

The United States and Norway are quite different concerning the question of whether it is necessary to establish a formal apparatus in close connection to the top political leader. In the United States, the executive political leadership is traditionally much more individually oriented and characterized by outsider politics—that is, it is extroverted, more entrepreneurial, conflictual (if necessary), taking positions, rhetorically oriented, and so on (Rockman 1991, 37–53). Norway, as a parliamentary system, has a more collectively oriented executive political leadership, characterized more by insider politics: consensus-building, consultation, and negotiation.

In the United States, the separation-of-powers system is obviously furthering such an apparatus; the same goes for the size and the general load in the system

(Kernell 1991, 192). This problem of capacity is believed also to influence other central political and administrative bodies. The president is separately elected and must have an apparatus to play on to fulfill the expectations of the voters. One important part of that is to have sufficient capacity to handle Congress; another part is both to control and give discretion to the federal bureaucracy at the same time (Moe 1990). But presidents in the United States have also generally mistrusted the federal bureaucracy, and in this perspective an apparatus close to the president is meant to counteract and control the federal bureaucracy—to create a checks-and-balances system inside the executive branch itself (Maidment and McGrew 1991, 76). This centralization, politicalization, and expansion of the presidency are believed also to be the effect of a responsive staff close to the president (Kernell 1989, 193–94; Moe 1985, 245–46).

In Norway, the need for such a formal instrument of prime ministerial power has not been apparent. The cooperation of the prime minister with the Storting is close, the coordination in the cabinet is effective, and the central bureaucracy is resourceful, politically controlled, and loyal. Therefore, the political and administrative capacity connected to the prime minister is limited and is a result of development in recent decades (Olsen 1983, 86). Unlike several other parliamentary regimes, Norway's government does not appear to be becoming more presidential.

The Executive Office of the President has expanded substantially during the previous 20 to 30 years (Kernell 1989, 188). Over a long period of time, it has developed gradually in complexity, becoming more specialized vertically and horizontally (Kernell 1991, 189; Maidment and McGrew 1991, 78–82). The structure of political advisers closest to the president has grown substantially, and the need for capacity for lawmaking, budget, economic oversight, and advice in domestic and foreign policies has gradually made the apparatus more structurally complex (Walcott and Hult 1987).

It is widely argued that this development of the Executive Office has resulted in an institutionalization of the presidency. Again, this characteristic had led many scholars to talk about the presidential branch—the *institutional presidency*—and the executive branch, the executive departments/agencies, and the independent regulatory agencies (Berman 1987, 99; Burke 1992; Hart 1987; Wyszomirski 1991, 86). The institutional presidency implies that the activities of these offices are generally influenced by rules and procedures. On the one hand, this institutionalization provides the president with an instrument characterized by greater stability and continuity, but on the other, it probably somewhat restricts the importance of personal features of each new president (Kernell 1989, 189).

Many scholars have been preoccupied with both the advantages and especially the problems with having a huge political-administrative apparatus close to the president, whether one is talking more generally about it or is concerned with certain organizational units (Hart 1987; Wildavsky 1987). One point of view is that the staff close to the president can have a life on its own, is overstaffed, has internal problems, and can shield the president from external information and contact with other important actors (Kernell 1989, 186, 235–36). Another view is that the president himself can strategically decide how much his staff should insulate him from the task environment and that he can manipulate the multiple organizations within his own apparatus and use the staff to handle complexity, ambiguity, and conflicts (Walcott and Hult 1987). In this way, the president can be said to both manipulate his staff and use it as an instrument, but potentially also to become a "prisoner" of it and have problems with its complexity. The greater the uncertainty, the more the staff of the president is differentiated, but also the greater the potential for specialized external interests to gain access to the office and its decisions (Hult and Walcott 1990).

In Norway, the prime minister's office for a long time consisted of only a few civil servants (Olsen 1983, 86). But during the past ten years the prime minister has added a limited number of political advisers and assistants, and the number of civil servants has also grown, although not by much. This slow build-up is obviously a reflection of the increased work load of the political leadership and is an attempt to strengthen coordination between ministries. The relatively limited size of the prime minister's apparatus indicates that many central political actors are skeptical about building up too much capacity, which relates both to the power of the Parliament and the capacity and expertise of the permanent bureaucracy.

Structure and Demography in the cabinet

The cabinet in the United States generally has a weak formal and actual political position, and it only rarely acts as a collective political body; the members serve more as advisers to the president (Maidment and McGrew 1991, 77; Martens 1979, 204; Rockman 1991, 37). It has never had a strong position in the president's mind, and that position has probably weakened over the years. One reason for this is that the many presidents have become skeptical about the clientelism in the executive branch and regard the heads of the executive departments as potential representatives of special interests. The members of the cabinet are in contact with the president on an individual basis, and this pattern of contact is biased in favor of the three most important members: the Secretaries of State, Defense, and Treasury.

The recruitment of the members of the cabinet is formally defined in such a way that members of Congress cannot be selected. cabinet members therefore tend to be "outsiders," from the business world or politicians who are old friends of the president. This means that the president, from time to time also an outsider, faces problems in using the members of the cabinet in relating to the

Congress.[1] This part of the demographic profile probably underlines the formal separation of powers and creates conflicts rather than cooperation with the legislature, quite the contrary in comparison to Norway.

But there is another feature of recruitment that may be operating in the opposite direction. This is especially true for the more specialized executive departments and agencies, with leaders who often are closely related to certain sectors and businesses, their geographical strongholds, and client groups (Peters 1989, 89; Riley 1987, 22). These are much more insiders in Washington, but the probability is that these leaders are not only building coalitions that help the president, but also are influenced by networks containing interest groups and committees and subcommittees in Congress.

The cabinet in Norway is formally defined as a relatively strong political body and generally acts very much on a collective basis. One indicator of this is that it meets more frequently and for longer hours than most other cabinets in Western Europe (Eriksen 1988a; Olsen 1983, 82). However, Norway is also experiencing the traditional conflict between ministers acting in the capacity of members of a collective cabinet and as political leaders of a ministry. It is typical that the cabinet and its members seem to balance this problem relatively easily, without excessively favoring special interests.

The formal difference in status and power between the head of government and the other ministers in Norway is much less than in the United States. But just as in the United States, there are differences between cabinet members. The prime minister is not paramount but is very important as a political organizer; the foreign minister and minister of finance are also important, and there is this slight inclination: the more specialized the ministry, the less status the minister has within the cabinet (Olsen 1983, 81).

There is a substantial difference between Norway and the United States concerning recruitment of the cabinet. In Norway, members of the Storting can also be members of the cabinet. When the prime minister is choosing the cabinet, he must balance the need for parliamentarian experience with the danger of weakening the party group(s) in Parliament too much—given that ministers must give up their parliamentary seats. Generally, about half the ministers have experience, and often from long service, in the Storting (Olsen 1983, 89). These effects of the recruitment profile seem to reinforce the effects of the constitutional structure—that is, it creates cooperation and close contacts between Parliament and the government. Not surprisingly, it also seems that the members of the cabinet who have parliamentary experience have the most political success. In Norway, ministers are also recruited from the interest groups for some of the more specialized ministries, but this background is often combined with and modified by experience with the party and the Parliament.

The Control of the Legislature by the Executive

The president of the United States not only has executive power, but also legislative powers. He has, over time, gained the power to initiate and shape the legislative agenda. But he can also veto legislation from Congress (Davidson 1988, 12):

> Once a bill or resolution has passed both houses of Congress and has been presented to the president, the president must sign or return it within 10 days, excluding Sundays. A two-thirds vote is required in each house to overrule a president's veto.

This executive veto makes the president of the United States different from most presidents in other democratic countries (Watson 1988, 37). In practice, relatively few of the many vetoes from the presidents are overridden (varying somewhat between presidents), showing that this is a potent political and legal weapon. The president also has the power of threatening the veto, thereby also influencing the lawmaking process and eventually changing the proposals. But many laws vetoed by the president are changed and proposed again, and then approved in different forms later (Watson 1988, 36).

Another way the president could potentially influence Congress with an instrument that has been discussed for a long time is through an item veto, which allows the president to veto parts of spending bills rather than the entire bills (Sundquist 1992, 281–93). Seen from the president's point of view, the reason for this instrument is to counteract public expenditures created by pork-barrel politics in Congress and to reduce the deficit. Such an instrument is of course controversial, since it would increase the president's influence relative to the Congress.

The item veto, or line-item veto, was used on a temporary and limited basis by the president with the consent of Congress before 1974, but was stopped through the Congressional Budget and Impoundment and Control Act. The president could then more indirectly increase his veto power through general lawmaking or constitutional amendments. In 1985, it was proposed formally in the Senate to establish a system with an item veto, on a temporary basis, but this proposal, like some others in the years to come, did not attain the needed majority for passage.

The constitutional separation-of-powers system in the United States defines a formal system of heterogeneity or pluralism. There are, however, different ways of looking at such a system, especially at how the president functions within it. Seen from the perspective of the president, it is difficult to use hierarchy and command in such a system, since it is designed to obstruct a president's attempting to centralize policy-making. Therefore the design of the system may lead to decision-making processes characterized by mutual adjustment, cooperation, negotiation, and the building of coalitions (Kernell 1991, 88–91). The president has some obvious advantages in such processes. Through his formal constitutional position the president has a broad perspective on political problems and a broad pattern of contacts, enabling him to build coalitions on an equally broad basis, both within and outside Congress.

Another perspective on the role and strategies of the president toward other actors, and especially Congress, is to emphasize conflicts in the system of separation of powers. A party theory of the presidency focuses on the importance of parties as teams, the conflict and struggle to win elections and define their results, the conflicts over public programs, institutions, and positions, the conflicts about building majorities, and so on (Kernell 1991, 91). In such processes, there are two major obstacles for a national party and the president as its actual leader: One is that there may be various degrees of agreement between the attitudes of the president and the constituencies, resulting in marked variation in loyalty from politicians of the same party in Congress. The other obstacle is the potential conflict between the president and Congress.

Observed from this point of view, the problems for the president relative to Congress appear to be less significant under a situation with a unified government. It will be much easier to use informal contacts and cooperation to coordinate preferences by consulting the majority in Congress from the same party (Kernell 1991, 95–96; Thurber 1991). There will always be some members of Congress from the president's party that are dissatisfied, even under unified government, but this seems to represent a somewhat less-significant problem.

Under divided government, however, party conflicts will dominate the relationship between the president and Congress. The actors will try consciously to design strategies and policies that are mutually obstructing, eventually resulting in the unfulfillment of their own goals (Kernell 1991, 97). This means that the president must use other strategies than the one used in a situation with unified government. He has to keep a distance from Congress, since it is difficult to compromise, and rely more on his own constitutional basis and resources, without cooperating closely with Congress. This strategy implies at least three elements: First, he can tighten the control over and politicize the bureaucracy, thereby preventing Congress from building "iron triangles." Second, he can strengthen his staff and especially the resources that are required for taking important issues directly to the voters and appeal to the public opinion over the heads of Congress (Davidson 1988, 12). Third, the president can use his veto powers, which is often believed to be his best weapon.

During the period between 1945 and 1990, a period dominated by divided government, the president used the veto 716 times, and his veto was overridden by a two-thirds majority only 6 percent of the time (Kernell 1991, 101). The use of the veto may, however, be problematic during some periods and in certain issues, because it is difficult to veto popular legislation or to create problems for forming coalitions between members of Congress from different parties. Therefore, the president can threaten to use the veto, but may better achieve his goals in some situations by negotiating with Congress.

There are different ways to define the formal constitutional power of the president. One distinction is between the *stewardship theory*, emphasizing the strong constitutional position of the president, and the *constructionist view*, stressing the constitutional limitations of the presidency related to the formal position and actions of Congress (Berman 1987, 55–56). But it is a fact that the Constitution does not clearly define the actual separation of powers, leaving room for conflict over the development of constitutional practice. Berman (1987, 62) emphasizes what he regards as an increasingly powerful president, having much to do with his position in foreign and security policies. This position over time encompasses the national emergency powers, the commander-in-chief position, treaty-making power, and the executive agreements and privilege. The judiciary and especially Congress have tried to limit the use and abuse of these powers, but they seem nonetheless to constitute a very important power basis for the president.

In the Norwegian political system, the Storting is formally in control of the executive power according to the parliamentary principle, not the other way around. But one can reformulate the question of executive control and ask in what ways the government can influence the Storting in public decision-making processes. The government has an enormous capacity to initiate and prepare decisions for the Storting. The cabinet has a large administrative apparatus to at its disposal and the Storting a relatively small one, even though it is gradually building up more capacity (Olsen 1983; Rommetvedt 1994). Some decisions—for example, law and formal rules—are often prepared partly by collegial bodies, committees, and commissions that have representatives from parties, interest groups, and the central administration, but the influence of the latter institution seems to be rather high (Christensen and Egeberg 1979, 1997a; Egeberg 1981).

However, there are also many reasons why we have to modify an argument of governmental dominance over the Storting, because of control over decision-preparation. First, the importance of stability, and the administrative apparatus as an agent of stability, is especially high between different cabinets. It is not easy for a new political regime to change priorities and the major structure of laws and the budget on a short-term basis. This is both an obstacle to democracy but also a safeguard, making life more predictable for citizens (Egeberg 1997; Jacobsen 1960).

Second, the preparation of decisions by the government is very seldom a conflict between "us and them." The government is based in one or more parties in the Storting and it/they, of course, participate in the organization of a law or budget process, thereby making it more difficult to divide the influence between the cabinet and Storting. Further, the informal contact between the government and Storting (including parties from the opposition), both politically and administratively, is so close and good that much mutual adjustment and "sounding out" is actually going on (Hernes and Nergaard 1990; Olsen 1983).

The Storting has also created counteracting instruments of control over the executive. These include a national audit office (the leader is an experienced representative, selected by the Storting), more information about different public institutions available through bulletins from the government, questions to the ministers in the Storting, meetings in the Storting with ministers and the undersecretaries of state, and so on, but the establishment of different ombudsmen is also relevant for control.

The government in Norway has a strong potential position of influence concerning the implementation of public policies decided upon in the Storting. The reason is a combination of solid expertise and decisions of a general character; for example, legal decisions that permit substantial discretion in the implementation process. But whether this is giving the government an independent, effective power base depends on many factors. One would be the existence of conflict, and normally most decisions to implement are not controversial. Another factor is that the resource base is often limited, so that the implementation process has to stay with certain limits.

A third factor is that implementing public policies demands cooperation of many other actors in addition to the central administrative apparatus. These would include the counties and local authorities, private organizations, and interest groups. The government has from this perspective only limited control over implementation, a fact reflected in problems of capacity of learning about the effects of public programs (Bratbak and Olsen 1980). A fourth factor is that Norway is a small and open society, thereby limiting the possibility for the government to take actions that are not approved by the Storting.

In summary, the government in Norway does not have the formal means of control over the legislature that the U.S. president has, but nonetheless is very important for the preparation of decision-premises and the implementation of public policy, but in a process of mutual adjustment and negotiation.

CONCLUSION

The president of the United States, on the one hand, has a solid electoral basis, a formal, powerful political position with many important functions, and a resourceful political-administrative apparatus to utilize. On the other hand, the president has many important constraints that undermine his powers; namely, the limitations of a weak party basis, the influence of Congress in organizing the federal bureaucracy and in appointments, the importance of the courts and due process, and so on. The prime minister of Norway has much less formal individual power than the U.S. president, but the collectively derived power is relatively higher through a strong party basis, a strong position in Parliament, and control over the organization of the central administration and appointments.

The internal structure of executive power in Norway and the United States is even more dissimilar than their legislatures: The U.S. president has developed a large organization, both in the general Executive Office of the President and more specifically in the White House Office (WHO). While the prime minister of Norway only has a small staff, resulting from great reluctance to give the top political leadership resources of coordination, adding to position power—historically the U.S. cabinet has not been politically significant, and the development seems more to have weakened than strengthened its position. In Norway, the cabinet has continuously been a very important, collective political body since the breakthrough of parliamentarianism in 1884. According to the macroconstitutional design and practice, the Norwegian cabinet has always been more based in the legislature in its recruitment as well as its powers than in the United States. This gives more consensus-oriented processes between the executive and legislative powers in Norway than in the United States, where the separate structural and demographic bases for the two powers create more conflicts and political struggle.

Note

1. The Clinton cabinet had some notable exceptions to this rule; for example, Les Aspin as Secretary of Defense and Ron Espy as Secretary of Agriculture.

CHAPTER FIVE

The Structure of the Executive Bureaucracy

WE HAVE STRESSED that the macrostructures of the political system in the United States, and the internal structures of Congress and the presidency, are much more heterogeneous and fragmented than the comparable structures in the Norwegian political-administrative system. This chapter will, based on this perspective, focus on whether this is also true for the central, executive bureaucracy in the two systems. Is the American federal bureaucracy primarily characterized by structural complexity and fragmentation (Seidman 1980)? Is this characteristic coupled with the multiple points of access for Congress and the bureaucracy as a battlefield between Congress and the president? Is the build-up and complexity of the presidency closely connected to problems of control over the federal bureaucracy? And, on the contrary, is the Norwegian central bureaucracy really characterized by structural homogeneity? Is this ultimately connected to the parliamentary principle, a close connection between the executive and legislature, strong parties in the Storting. and the emphasis placed on coordination and control from the political leadership?

The chapter begins with a historical overview of the development of the central executive bureaucracy in the two systems, followed by a contemporary comparison of the internal structure in the two administrative systems, before finishing with a short discussion about the executive political leaders' control of the executive bureaucracy. This analysis will demonstrate the extent to which these administrative structures are related to the patterns already documented for other parts of the two political systems.

The Historical Development of Public Administration in the United States

The First Seventy Years

In 1789, when George Washington began to organize his first administration under the new Constitution, he inherited a tiny administrative apparatus from the earlier government, operating under the Articles of Confederation. There were only a few clerks in the War Department, the Treasury, and in the Post Office. There was also a small dispersed field staff, consisting mostly of employees of the Post Office and some Customs officers, attempting to protect the already-extensive borders of the thirteen states against smugglers (White 1948). The conditions that President Washington inherited were in some ways indicative of the future for public administration in the United States, given that the clerks' salaries were far in arrears and there was little faith in the capacity of government to perform its assigned tasks. However, from this modest beginning a large and professionalized public bureaucracy has been created, even if most Americans would not willingly acknowledge its abilities and accomplishments.

Washington's cabinet was composed of only four officials: the Secretaries of War, State, Treasury, and the Attorney General. At its inception, the American government was very much confined to the "defining functions" of government (Rose 1976): Protecting the population against internal and domestic threats, providing overseas representation, and collecting taxes. To the extent that government performed other tasks, they were performed by the states and localities with almost no federal concern or intervention. Thus, much of the early involvement of government in economic development and social programs (especially education) was the responsibility of state and local governments. In practice, that involvement was substantial and despite the mythology, economic development in the United States had substantial public-sector involvement from the beginning of the Republic (Hughes 1991; Sbragia 1996).

Although it was not directly involved in most of this public-sector activity, the federal government even then was beginning to play a role in a wider range of policy areas than might be thought from looking at the formal organization chart. Of greatest importance was the Northwest Ordinance of 1787. This was one of the first major pieces of legislation of central government in the United States (actually passed under the Articles of Confederation), and it provided for the organization of vast stretches of public land in what is now the Midwest. The Ordinance applied to all parts of the states of Illinois, Indiana, Iowa, Michigan, Minnesota, Ohio, and Wisconsin. The model of land division and local government was extended to several other Western states. This vast amount of land was surveyed and divided systematically under federal direction, with requirements for some land to be set aside for public education in each surveyed section. Further, largely for military reasons, the Army Corps of Engineers and the Navy began a series of infrastructure developments (at least from 1824 onward) that were to be important for subsequent economic development.

Although employment in this small federal government fluctuated significantly, usually in response to wars, its basic scope of activity changed very little for most of the first one hundred years of its existence (Peters 1985). Even at the time of the Civil War (1861–1865), the federal government had added only a single cabinet department—Interior, in 1849. The federal government had, however, begun to create a variety of independent executive agencies and bureaus that would eventually evolve into executive departments. For example, the foundation for the future Department of Agriculture had already been established by such independent organizations as the Biological Survey, formed during the 1830s.

Even at this time some of the basic structure of U.S. cabinet departments was being established, with the components of the larger organizations beginning to act somewhat autonomously and developing their own relationships with Congress. For example, the Department of Interior was composed of several large agencies, most notably the General Land Office and the Office of Indian Affairs. Those organizations had substantial constituencies inside and outside of Congress and could operate somewhat independently. Certainly, at the time of the Civil War, President Lincoln could not be sure that his departments were indeed his and were not more in the service of the radical Republicans in Congress or the clients of the departments (White 1958).

Following from the above discussion, we will be pointing out that perhaps the most important feature of U.S. public administration is the relative autonomy of the agencies existing within the cabinet departments. Unlike the independent bodies in Norway, this autonomy does not have any constitutional or legal status, but rather has developed more for political reasons. In addition, all through its history, Congress has created a number of independent executive agencies outside the cabinet departments. The General Land Office and the Biological Survey, for example, were created as independent executive agencies. More recently, important organizations such as the General Services Agency, NASA, and the Environmental Protection Agency have been created in the same format. Even more than the agencies located within the departments, these independent agencies must rely upon their own political connections, both with the president and with congressional committees, for survival and for continued funding for their programs.

Public employment may have been relatively small during this first century of the Republic, but positions in that nascent bureaucracy were valued. There was a relatively small, professional public service, with the vast majority of public jobs going to political appointees. The Jacksonian conception of the American political system was that almost any citizen was qualified for almost any public job, and therefore that there was little justification for a professional public service (White 1954). Indeed, such a professionalized government was considered undemocratic in this populist conception of governing. Therefore the "spoils system" was introduced to fill most public jobs, with the president, or his higher appointees such as cabinet secretaries, being given the opportunity to make numerous appointments, some without approval of the Senate (Hoogenboom 1968).

As a consequence of the spoils system, there was significant turnover in employees when there was a change in top officials. This pattern of politicized change in officials has been continued for a larger proportion of positions in the United States than in other industrialized democracies (Mackenzie 1987; Neustadt 1960). Some 3,000 positions are available for appointment every time there is a change in administration, and that number has been tending to increase with each administration after the Eisenhower years (Light 1996). In particular, the Civil Service Reform Act of 1978 permitted up to 10 percent of the newly created Senior Executive Service (the former "supergrades") to be selected by political appointment. This reliance on political appointments and attempts at direct political control over the bureaucracy does, however, have deep historical roots in American political life.

The Republican Era

The first major period of expansion of the federal government in the United States was during the era of domination by the Republican party, roughly from 1860 to the election of Franklin D. Roosevelt in 1932. This period of hegemony was broken only by the two elections of Grover Cleveland, in 1885 and again in 1893, and by the election of Woodrow Wilson in 1912, but the Republican party was dominant. This was also the period of very rapid economic and geographical expansion for the United States, and those developments created the need for a larger and more active federal government. One cabinet department, Agriculture, was created during the Lincoln administration; although titled the "Department of Agriculture," it did not achieve cabinet status until the 1890s. Until then, it was headed by a Commissioner of Agriculture rather than a Secretary of Agriculture. Another department, Labor and Commerce, was created around the turn of the century. It was split into the two separate departments of Commerce and Labor early in the Wilson administration (1913).

Both of the new cabinet departments were responses to the increasing role of the federal government in American economic life. The development of a variety of economic regulatory organizations even more clearly indicated the necessary role of government regulation for the emerging corporate economy. The Interstate Commerce Commission (ICC) was created in 1887 with the principal task of regulating rates and other aspects of railroad activities. The railroads were crucial to economic success for all parts of the economy, including agriculture, at that time. The numerous abuses of that economic power by the railroads, and the growing populist movement, helped produce this new organizational form of federal intervention that would subsequently serve as the model for a number of other federal regulatory interventions.

As well as being a model for policy, the Interstate Commerce Commission also became an organizational model.[1] One of the dominant ideas of the populists was that politics was a major impediment to effective governance. As a consequence, the ICC was created as an independent, depoliticized organization, with the president having a role only in the initial appointment of its commissioners. Those officials were to serve long terms and be protected in so far as possible from political influences. Further, no more than three members of the ICC could come from the same political party, so that even if one party controlled the White House for a long period—for example, the twenty years of Democrat rule from 1933 to 1953—it could not totally pack the Commission.

While this design was well intentioned, it also proved to be largely unsuccessful. It is almost impossible to insulate any public organization from politics, so the only question is how that politics is to be played out. In the case of the independent regulatory commissions, the politics have been interest-group politics, and this has most often meant their "capture" by the very interests they were intended to regulate (Calvert, McCubbins, and Weingast 1989; Huntington 1952). Not surprisingly, some of the major opposition to deregulation in the United States during the past several decades has come from the regulated industries, who did not want to have to compete in an open market.

The use of independent regulatory commissions was expanded during the first part of the twentieth century. One of the most important of these commissions was the Federal Trade Commission, created in 1914. This was another reaction to the corporate economy, and especially the growth of economic concentrations in monopolies and trusts. First through the Sherman Act (1890), administered by the Department of Justice, and then through the Federal Trade Commission Act and the Clayton Act (1914), the federal government attempted to restrain these activities "in restraint of trade" (Eisner 1993). The Federal Trade Commission was designed to regulate not only trusts but also other uncompetitive acts such as deceptive advertising. This commission was then followed by other commissions such as the Federal Communications Commission, the Federal Power Commission, and (much later) the Nuclear Regulatory Commission—all designed to regulate various segments of the economy.

The second great creation of this period (for our purposes) was the Pendleton Act, and with it came the beginnings of the civil service system. President James Garfield was assassinated in 1881 by a disappointed office seeker, and his death served as the catalyst for instituting a merit system for employment in the federal government. Although the assassination was the immediate impetus, the move to a merit system for public employment also conformed with the populist thinking of the time and belief that expertise was more important than political commitment in making public policy. This sentiment was expressed very clearly by (future president) Woodrow Wilson in his famous (1887) essay on government. Also, the Pendleton Act was the culmination of a number of attempts to initiate a merit system in the United States, the previous ones being defeated when faced with resistance from entrenched political forces dependent on patronage (Johnson and Libecap 1994, 30–33).

Although the Pendleton Act was the beginning of the merit system, it was a very modest beginning. Only half of total federal employment was covered by merit appointment by 1904, and these were primarily low-level clerical positions and the Post Office. These positions were supplemented by additional positions in following years, often through the dynamic of "blanketing in" political appointees—that is, a president would make his political appointments to positions and then convert them to merit positions so that they could not be removed by the next president. By the time of World War I, approximately 60 percent of total federal civilian employment was by merit, and that number continued to increase, albeit slowly.

The third aspect of American government life that was changing during this period was that the foundations were being laid for the emergence of its own version of a social-welfare state. One part of this was the emergence of labor as a significant political and economic force, and with it the beginning of some labor programs and regulations on the uses of labor. Also, the federal government began to play a role in education, in part through the passage of the Morrill Act (1863) to fund the land-grant colleges that have been so important for American higher education. Finally, the federal government began to provide (extremely limited) social supports for some segments of society, although the states continued to be the major actors in this area.

The Democratic Era

The Great Depression beginning in 1929 was the root of a fundamental transformation of American government. Franklin D. Roosevelt was elected president in 1932 and began a period of rapid and adventurous organizational innovation. There were several features of that transformation. Perhaps the most crucial was the development of a rudimentary but still real welfare state. The cornerstone of that effort was the Social Security Act of 1935. This, however, was not the only important social program of the period. Others like the Farm Security Administration provided benefits to rural America, and several employment programs, such as the Civilian Conservation Corps and Works Progress Administration, were (thinly) disguised welfare programs.

Another trend of the Roosevelt administration was the increasing use of independent executive agencies to address important policy problems. For example, organizations such as the Works Progress Administration and the Public Works Administration were established to provide jobs and rebuild the economic infrastructure of the country. Franklin Roosevelt had a tremendous capacity for work and an equally great need for information, so he organized these agencies under his personal control (Burns 1956). The leaders of the organizations, as well as a number of other officials, would be appointed by the president and would be expected to be his eyes and ears in the organization.

Besides the number of public organizations created, the Roosevelt administration also experimented with a variety of organizational formats. For example, the Tennessee Valley Authority, which has since become the largest single electric utility in the United States, was organized as an independent public corporation (Hargrove 1994; Selznick 1949). The National Recovery Administration and the Agricultural Adjustment Administration were organized along corporatist lines, with a pronounced role of private-sector organizations in making and implementing public policy (Himmelberg 1994). The Roosevelt administration also witnessed the beginnings of an independent White House staff in the service of the president. For example, the Bureau of the Budget was created to provide the president with help in managing the rapidly growing federal budget (Savoie 1995).

In addition to the experiments with organizational formats, the Roosevelt administration also created somewhat greater respectability for working in the public sector. The civil service was accorded a more significant position in American life and the "best and brightest" were attracted to Washington to help solve the problems of the Depression and to help win the war. This was one of the few times in American history in which a premium was placed on public service, and a number of people who were to have long and distinguished careers in government were recruited at this time. There were few significant changes in the structure of the civil service, but a group of distinguished individuals were brought into the public sector.

Another aspect of the Roosevelt administration's organizational developments was that they served as the foundation for the continuing development of the welfare state. Indeed, the launching of the Social Security program in 1935 was the principal milestone in American social policy. That program was crucial, but it was not the only social program launched during the period. The Farm Security Administration attacked problems of rural poverty and although it had little demonstrable effect at the time, it did have pronounced long-term effects (Salamon 1981). In addition, this program with its wide range of services can be seen as a model for some of the programs in President Lyndon Johnson's War on Poverty. The Roosevelt administration also made some of initial steps in the creation of a program of public housing for less-affluent Americans.

The creation of a number of relatively autonomous organizations during the Roosevelt years was another era in this type of organizational development in the United States. Many of these organizations were short-lived: one lasted less than a month before being consolidated into another organization. This activity indicated some belief that the answer to the problems of government could be found by getting the machinery of government right. This has been a consistent theme in American government, but was never more operative than during the New Deal. That theme was reinforced by the theorizing of the time, which assumed that objective, scientific laws of administration could be uncovered (Gulick and Urwick 1937).

Finally, the last few years of the Roosevelt administration witnessed a massive expansion of the federal government to meet the demands of World War II. Federal civilian employment expanded, and this was accompanied by another outburst of organizational creativity. Most of the new organizations created were in familiar models (e.g., agencies of the War Department, the Commerce Department, or the Department of Agriculture), but the purposes to which they were applied were novel for the country. For example, organizations such as the Office of Price Administration and the Office of Economic Stabilization were responsible for managing most of the economy. They were able to gain this much power given the crisis created by the war, although it was widely agreed that this was to be a short-term deviation from the American commitment to free enterprise.

Although the direct intervention in the economy through planning and direct controls was definitely a short-term program because of the war, this period did substantiate the legitimacy of government action in the economy in the United States. The shift in thinking about government and the economy was manifested in the passage of the Full Employment Act in 1946. This Act represented the American acceptance of Keynesian economics and the belief that government could effectively manage the economy (Hall 1989). Further, with the creation of the Council of Economic Advisors, economic management in addition to the earlier creation of the Bureau of the Budget had become very clearly a presidential responsibility. The experiments with government control during the war were then the beginning of a much larger and more interventionist public sector.

Another act of utmost importance for public administration occurred during the period immediately following World War II. Reflecting the experience of a government that had been growing with relatively little control or consistency, the Administrative Procedures Act of 1946 (APA) was a major attempt to codify administrative procedures for the federal government. In particular, the APA established rules for writing secondary legislation ("regulations" in the language of American government) that mandated the opportunity for citizens to have some participation in writing that legislation (Freedman 1980). In addition, the APA specified the procedures for administrative adjudication, a form of adjudication that was far outstripping the federal courts in terms of the number of cases processed each year (Mashaw 1983).

The two elections of Dwight Eisenhower as president can be seen as either an interlude in the period of Democratic control or as the beginning of a long period of Republican domination of the presidency. Another way to think about these elections was as being the beginning of a long period of divided government (Fiorina 1992; Sundquist 1988). Whatever the characterization, there was relatively little change from the nature of the state created during the Roosevelt years. If anything, government expanded during the Eisenhower years in large part in response to the continuing defense commitments of the Cold War. A number of different functions were added to the federal government, using the rubric of defense for programs such as the Interstate Highway System and the National Defense Education Act.

The growth of government in response to the Cold War during the Eisenhower administration followed a major reorganization of the defense function during the Truman administration (Neu 1987). The National Security Act Amendments of 1949 created a (more) unified defense establishment with the former War Department[2] and Navy Department reorganized as components of a single Department of Defense, along with a new Department of the Air Force as an equal organization. In addition, a number of common functions for the military services and defense management was consolidated into organizations such as the National Security Administration. Finally, the National Security Council and the Central Intelligence Agency were created as a part of the same legislation in order to integrate intelligence collection and to provide the president with sources of policy advice on defense and foreign affairs independent of the Departments of Defense and State.

During the Eisenhower years the government also became more structurally complex. The Eisenhower years were a time of great organizational activity in government, with a number of programs eliminated but also a number of programs created (Hogwood and Peters 1983). Several cabinet-level departments were created or enhanced during that time, most importantly the Department of Health, Education, and Welfare (DHEW). Other major organizational creations of this time period were the National Aeronautics and Space Administration (NASA) and the Small Business Administration. At the same time that these organizations were being developed, a great deal of the administrative infrastructure developed to fight World War II and then the Korean War was abolished, along with some of the remnants of the New Deal. The social safety net initiated during the Roosevelt years was not, however, loosened, and if anything, was expanded through the creation of DHEW and increasing policy activism by its leadership.

The Great Society Programs and Beyond

The contemporary American bureaucracy is the product of accretion over a number of years, but it was affected profoundly by the Great Society programs of President Lyndon Johnson. After the assassination of John F. Kennedy, President Johnson had a great deal of political capital with which to attempt to transform American society through public action (Kaplan 1986). He also had a great deal of personal ability to play the political games necessary to have his legislation enacted by Congress (Caro 1982). The combination of these features produced a flowering of government programs and associated organizations. Although some of these programs were organized within existing cabinet departments, many were also organized outside those organizations, in order to provide them with greater freedom and greater capacity to innovate. For example, the Office of Economic Opportunity was created outside of DHEW and the Department of Labor, although many of its social programs could well have been located in those departments.

In addition to creating organizations within government, the War on Poverty and the Great Society also fostered closer links between the society and government. Following from some earlier urban-renewal programs, many Great Society programs required the "maximal feasible participation" of the people affected by the programs (but see Moynihan 1972). In many cases this participation required the creation of organizations among poor people where none had existed. Although critics considered that these participatory requirements may have been counterproductive in the end, they did begin to create a more participatory culture among disadvantaged and effectively disenfranchised elements of the population.

Given the generally very conservative image of Richard Nixon, we might not expect any significant administrative changes other than perhaps reductions in force during his administration, but that was far from the case. Two very important organizational changes occurred during this time. One was the creation of the Office of Management and Budget (OMB), from the older Bureau of the Budget. Still located in the Executive Office of the President, OMB was given the additional responsibility for improving management in the federal government. This gave it a wider scope of authority and enabled the president to become involved in issues that might have escaped his attention before the reorganization of this very important "central agency" (Hart 1994). The other change of great importance during the Nixon administration was the creation of the Environmental Protection Agency (EPA), as an independent executive agency. Although certainly President Nixon did not foresee the future expansion of this organization and its influence in the economy, the movement into what was largely a new area of federal concern, albeit in partnership with the states, was supported by his administration.

The above two positive changes can be contrasted with the growing politicization of the federal civil service during the Nixon administration. In his attempt to create an "administrative presidency" (Nathan 1975), the president and his associates sought to ensure the commitment of the civil service to the program of the presidency. This was almost impossible to do for the career civil service, especially those who had been recruited during a period of more activist government (Aberbach and Rockman 1976), but it was possible to try. The "Malek Manual" provided managers loyal to the president (usually political appointees) with advice about how to make the lives of recalcitrant civil servants miserable, so that they would leave government. Further, the number of political appointees was increased so that there could be even greater control. American government has always had a many more opportunities for political appointments than is true for European governments (see Meyer 1985), but these became even more important during the Nixon years.

The excesses of the Nixon years were followed by attempts by a technocratic Democratic president, Jimmy Carter, to both clean up the politicization in government and to create a more effective and efficient civil service.[3] The major manifestation of this commitment to reform was the Civil Service Reform Act of 1978 (CSRA). This Act contained a number of provisions that continue to influence the conduct of the civil service. First, the Act divided the Civil Service Commission into two parts: the Office of Personnel Management (OPM) and the Merit Systems Protection Board. The former office was responsible for managing the civil service of the United States, while the second had more of a quasi-judiciary function to regulate the implementation of merit system laws and to ensure that OPM and other federal personnel offices were complying with the law. This was consistent with other reforms of the Carter administration that separated administrative from regulatory activities of federal organizations; for example, dividing the Atomic Energy Commission into the Energy Research and Development Administration and the Nuclear Regulatory Commission.

The Civil Service Reform Act also created the Senior Executive Service (SES) out of the former "supergrades" in the civil service (Ingraham and Ban 1984). The idea of the SES was to create within American government a cadre of career public managers that could move from position to position, with the rank remaining with the person rather than with the particular position held at the time. Further, these officials were available to be moved around within government to meet the needs of government, rather than remaining in one agency for most of their careers. These SES members were also to be subject to greater performance measurement that could produce either benefits (bonuses for outstanding performance) or sanctions (potential dismissal for poor performance). A merit-pay system was also planned for middle managers under this Act. The real results of this legislation continue to be debated in and out of government (Ingraham and Rosenbloom 1992).

In addition to the reform of the civil service, the Carter administration also engaged in other efforts at reorganization (Szanton 1981). Its efforts included the consolidation of a number of independent administrative and regulatory organizations and parts of other departments into cabinet-level Departments of Energy and of Transportation, and the splitting of the Department of Health, Education, and Welfare into the Department of Education and the Department of Health and Human Resources. In addition, there were a number of organizational changes at the subdepartmental level, all attempting to enhance the efficiency and responsibility of the federal government.

Another important administrative change during the Carter administration was the adoption of the Inspectors General Act. Under the terms of this Act, an office of the Inspector General (IG) has been created in each cabinet department and large executive agency. This office was expected to root out "fraud, waste and abuse" in government organizations and to use its investigatory powers to attempt to make those organizations more efficient and more accountable (Light 1993; Morris and Gates 1986). The offices of the IG were also intended to make it easier for employees to report any malfeasance on the part of the organization or of their organizational superiors—it made it easier for them to "blow the whistle."

The Reagan Revolution

The election of Ronald Reagan as president in 1980 ushered in a new era for government and the civil service. A number of presidents, including Jimmy Carter, had been elected by running against Washington and promising to reform the civil service. The Reagan administration differed both in the radical nature of the reforms proposed and in the dedication to implement those "reforms." The reforms actually implemented were not as radial as proposed: the Department of Education was not abolished, and very few other programs were actually abolished. Still, some significant changes were produced during the eight years of this administration.

Some reforms of the public sector under Reagan were relatively benign and widely accepted throughout the public sector. For example, the president's Management Improvement Council and most of the components of Reform '88 contained a number of changes, such as better cash management, improved purchasing procedures and facilities management, that simply improved the administrative performance of government. If anything, these changes tended to enhance the position of the career civil service because they made it possible for the managers to manage. Other reforms were not so benign in the eyes of the career civil service and were resisted by it whenever possible. One of the more obvious of these reforms was the Grace Commission, which brought over 1,200 private-sector executives to Washington to analyze the operations of government and recommend changes to promote greater efficiency. Few of their over 2,400 recommendations were ever actually implemented, but they did create a great deal of continuing animosity between the Reagan administration and its civil servants.

Also, the Reagan administration took the provisions of the Civil Service Reform Act ("Carter's gift to Reagan") and used them fully to politicize the public service even further than it always has been (Peters 1986). The CSRA permitted up to 10 percent of the appointments in the general Senior Executive Service to be political rather than career oriented.[4] The Reagan administration exercised this option to its fullest, and also applied a "political litmus test" to each of its appointments to ensure that all the appointees agreed with the program of the president. Even when career civil servants were retained in important positions, they were usually kept "out of the loop" for significant decisions. This required more political appointees and was a major part of the "thickening" of government (Light 1996).

Although not really identifiable as a program, a number of other decisions of the Reagan administration tended to diminish the standing of the career civil service. One of these actions was to reduce the real compensation of civil servants. First, although the president's Pay Agent and the Advisory Commission on Federal Pay[5] both recommended substantial pay increases, especially for political and career managers, the president recommended much lower pay increases (Peters 1985). Likewise, the Civil Service Reform Act of 1978 recommended substantial funding for bonuses for SES members, but these were never funded adequately by the president and Congress. While the civil service had never been a good career in which to get rich, it became even less rewarding during the Reagan years (King and Peters 1994).

The Reagan administration also publicly and privately denigrated the role and quality of the American civil service (Benda and Levine 1986). For example, one (politically appointed) official argued that civil servants did not have to be paid well, because if they were really capable they would move to well-paying jobs in the private sector (Levine and Kleman 1993). Likewise, another official said that civil servants needed only to be "competent" and that the public sector should not strive for excellence (Comarow 1981); the most qualified individuals should work in the private sector. In short, the Reagan administration tended to regard the civil servant as at best a necessary evil, and at worst a real impediment to the success of the federal government.

The Clinton Administration

After the rather uneventful Bush administration,[6] the Clinton presidency has returned to greater activism in reforming the civil service, and the public sector more generally. The most important effort in this direction has been the National Performance Review (1993), otherwise known as the Gore Commission after its head, Vice President Al Gore. Unlike many other reform efforts in the United States, the Gore Commission was formed primarily of civil servants rather than outsiders from the private sector (Peters and Savoie 1994). A few outsiders were brought in, but were mostly from state and local government and were therefore familiar with the constraints under which governments operate. These civil servants were given only a short time period to recommend changes across the gamut of federal government organizations that would produce a more effective and "user-friendly" government.

The guiding principles for the proposals of the Gore Report came from Osborne and Gaebler's (1992) ideas about "reinventing government." As one part of the reinvention exercise, approximately one-seventh of federal civil service positions were to be eliminated before the year 2000. Most of the positions eliminated would be at middle-management levels, a reduction made possible by reducing the degree of internal supervision and regulation in the federal government in line with the idea of empowering lower echelon workers in public organizations. After the Republican victory in the 1994 elections, these reductions appeared meager, so "REGO2" in the FY-1996 budget proposals called for considering even sharper reductions in the civil service.

The Gore Report did not provide any uniform template to use throughout government, but rather adopted a somewhat more experimental model of change. Each cabinet department and large agency was to develop its own "reinvention laboratory" that would be able to experiment with various approaches to improving government. The small parts of the Gore Commission remaining in existence will monitor these laboratories and attempt to diffuse the successful reform ideas throughout the rest of the federal government. The Gore Report did not promise instant results, but instead projected that at least ten years would be required to create the leaner and more effective government it was seeking.

In addition to the commitment of the Clinton administration to these administrative reforms, there was an attempt to create a "virtual organization" that would foster and maintain interest within the civil service itself (Peters and Savoie 1994). This type of linkage also developed among the "reinvention laboratories" created as a result of the Gore effort.

Not surprisingly, reforms of this type are not popular with everyone in government nor with all academic analysts of the public sector (Moe 1994). On the one hand, the critics argue that the reforms being proposed are not really new but merely represent the continuation of trends in administrative reform that were already under way well before the Gore Commission was created. Further, it has been argued that the emphasis of management tends to devalue other aspects of the role of civil servants; for example, the role in policy advice. Further, the managerial focus tends to make it appear that the real problems of government are the fault of the bureaucracy, and if government would only adopt good (private-sector) management techniques, then all the problems would be solved.[7] The critics tend to think that policy problems are much more complex than that and what is really needed is a more thorough analysis of policy and greater emphasis on traditional administrative values, such as accountability.

The Republican triumph in the congressional elections of 1994 ushered in an era of apparently even greater skepticism about government and the role of the civil service. While the "Contract with America" that served as a major part of their campaign does not contain any specific mention of the civil service, it does have several provisions that can be expected to place the role of the civil service into even greater doubt than it usually is in the United States. First, there is a demand for a reduction in the ability of the bureaucracy to issue new regulations. Most laws of Congress are rather vague and require elaboration by the administrative agencies responsible for their implementation before they can have any real effect (Kerwin 1994). The reforms proposed would require more economic analysis of the proposed regulations and greater ease in blocking any new regulations (McGarrity 1991).

The second aspect of the Republican contract, relevant for public administration, is the set of mechanisms designed to reduce public expenditure and to eliminate public programs. While these provisions are not *per se* an attack on the position and perquisites of the federal civil service, they do not auger well for the careers of civil servants. These proposed budget cuts, combined with the exercise in reducing employment already well underway as a result of the Gore Commission, would result in a significant reduction of positions and with that a diminution of career prospects for many civil servants. Civil servants historically have selected government as a career because of the apparent security of civil service jobs, but security certainly does not appear to be a realistic attraction any longer. Of course, all of these proposed employment cuts may not be implemented and government organizations have a variety of mechanisms with which to protect their interests, but still there is less certainty about what government will look like in the near future.

THE HISTORICAL DEVELOPMENT OF THE CENTRAL ADMINISTRATION IN NORWAY

The structure of the central Norwegian executive bureaucracy, or central administration, is and has always been quite simple, compared to the United States, both concerning inter- and intraorganizational features. For most of the time since 1814 there have been two main types of administrative bodies in the central administration, the ministries and the directorates (executive agency-like). The ministries are, and have since 1814 mainly been, hierarchical organizations just like the directorates, with the directorates organized outside and hierarchically subordinated to the ministries.

The Development of Ministries: Structure, Demography, and Behavior

Structural Development: Horizontal and Vertical Specialization. The ministries (executive departments) have been the superior, political-administrative bodies in Norway since 1814–1815. They replaced a central administrative structure based on a collegial model, demonstrating how certain organizational structures or forms are popular during certain periods (Meyer and Rowan 1977). There were at first six ministries, all recognizable in today's structure (Maurseth 1979): Ministry of Church and Education, Ministry of Justice, Ministry of Police (now together with Justice), Ministry of the Interior, Ministry of Commerce and Finance, and a War Ministry. During the next 130 years the number of ministries increased only very slowly, compared to a more marked increase after World War II.

There have been two waves of creating new ministries during the last forty years (Christensen and Egeberg 1997a). One wave occurred right after World War II with the establishment of ministries for fisheries, commerce and shipping, industry, local government and labor, and communications, a combination of old ministries in new clothes and ministries organized around new policy areas. The last establishments in this wave were ministries for consumers/families and wages/prices during the mid-1950s. The second wave occurred in the 1970s and 1980s, with ministries for administrative development, environment, petroleum and energy, culture, and science, and for help to underdeveloped countries. This wave typically encompassed new policy areas now being given organizational attention on a regular basis.

During the last five years we have seen a certain rationalization in the ministerial structure, with the fusion of the industry and petroleum/energy ministries, as well as commerce, underdeveloped countries, and foreign affairs. One interesting feature is that Norway now for the first time has more than one minister in some ministries (foreign affairs and health/social affairs), a development creating both increased political capacity and problems of the allocation of authority. This is not a system using junior ministers, but rather one with formally equal ministers having authority over different parts of a ministry. In 1996 the Labor Party selected a new prime minister and also changed some of the ministerial structure in a more specialized direction again. Today, Norway has sixteen ministries (with eighteen ministers) and the prime minister's office.

Between 1814 and the present, Norway has experienced a substantial increase in the number of people working in the ministries, even though compared to other countries Norway does not score high. In 1814 only sixty people worked in the ministries, while 170 years later there were only 3,150. The development mentioned above implies a long period of increased interorganizational, horizontal ministerial specialization, even though lately there has been some despecialization. But the ministries also have experienced an enormous increase in their internal complexity since 1814. As a measure of internal horizontal specialization, there were 32 divisions in the ministries all together in 1844, while the figures were 353 in 1982 (Laegreid and Roness 1983, 12). In 1914 there were a total of sixteen departments inside the ministries, while in 1982 there were ninety-three.

After World War II there were waves of new types of departments and divisions inside the ministries, such as economy and administration, planning, international units, information units, administrative development, and so on (Christensen and Egeberg 1997a). The ministries have lately attempted to loosen up some of the strict formal division of labor by allowing for a more flexible matrix-like structure in some parts, greater use of collegial groups, and greater use of sections, but the main format is still a hierarchical structure.

During the last fifty years there has also been a sharp increase in the degree of internal, vertical specialization in the ministries, with an increase in the number of hierarchical levels. The new types of leadership positions include secretary general—the administrative leader in a ministry—and assistant director-general and deputy director-general. The latter two offices are leadership positions created between the director-general (heading the department) and the head of the divisions as a means of enhancing internal coordination.

Of greater importance for some of the top leadership positions in the ministries is the creation of a number of political helpers for the ministers. For a long time, ministers were the only political appointees in the ministries, but after World War II there has been a steady growth in the number of undersecretaries of state and other types of political advisors. There are now approximately sixty or seventy people all together in these positions, not, comparatively, a very impressive figure. Some undersecretaries of state may take over some parts of ministers' tasks or have a quasi-administrative profile, creating conflicts of domains of influence with the directors-general (Eriksen 1988b).

About ten or twenty years ago some Norwegian ministries started out by establishing positions as advisors or special advisors, motivated partly by creating alternative career paths with increased wages, partly by a desire to prevent some civil servants from being promoted to leaders (Christensen and Egeberg 1997a). This development, encompassing a growing number of civil servants during the last decade, has increased motivation, but also created more ambiguity concerning authority, leading to efforts to couple advisors more clearly again with the line organization in the ministries (Eriksen 1988b).

Thus, Norwegian ministries have increased much in complexity over time, especially after World War II. This complexity has been one important way of coping with increased problems of capacity, created by the increased load and complexity of the public agenda and more general laws that then leave greater discretion in the hands of civil servants.

The increased importance of interest groups in an integrated organizational participation in Norwegian public policy is a reflection of problems of capacity for the political and administrative leaders, but also the increased complexity of the administrative apparatus and public policies (Christensen and Egeberg 1997b; Olsen 1983). The more complexity, the more points of access that exist for the groups. The interest groups are legitimate participants as affected parties, experts, and important actors for the implementation of public policies (Egeberg 1981). On the one hand, the ministries are constrained by their contacts with the interest groups; on the other hand, they can also build political alliances as well as enhance their influence over the groups. The interest groups are participating both as affected "outsiders" and "insiders." The latter factor related to their organization of employees in the governmental organizations.

Characteristics of Personnel Policy. The institutional, personnel policy toward the civil servants in the central administration seems to be divided into four types: internal, personnel administrative policy; wage policy; the policy of participation; and the recruitment or career policy (Laegreid 1989). Historically, the decision-making processes of the first three of these elements have been characterized by a combination of central control by the government and negotiations with the peak organizations in the public sector, especially the Norwegian Federation of Trade Unions.

The policy directed towards internal, personnel administration has historically been centralized and coordinated by the government, but developed more into negotiations over time. The net result of this development has been an institutionalization of a very complicated law for civil servants, with a gradually stronger emphasis on their rights and rule of law.

The wage policy has always been closely coupled to the personnel-administrative policy, and formally organized together in the Ministry of Planning and Coordination (known as the Ministry of Government Administration until the fall of 1996). In negotiations between the government and the peak labor organizations for civil servants, an institutionalized unambiguous norm about what principles the wage system should be based on has been adopted: standardization, equality, tenure, education, and so on. The tightly controlled and coordinated wage policy has been loosened up somewhat during the last ten years, allowing merit pay for certain leaders and experts, but it still pays very strict attention to the traditional values such as wage solidarity (Laegreid 1993a, b).

Historically, the policy towards the participation of civil servants in the "internal democracy" has been characterized by variety and attention to different norms and values in different parts of the apparatus. Since the 1970s there has been more emphasis on the rights of the organizations than on individual rights. One driving force behind this emphasis was the demands from the peak organizations to formalize and coordinate this policy more, and to weaken the influence of the government. Since then, the policy has been more standardized and coordinated, but the political-administrative leadership still has substantial control, although the participation rights for the civil servants have been widened. If the leaders define a process as politically important—for example, concerning reorganization or recruitment—they can decide that the decision-making process should be relatively closed and without too much influence from the organizations of the civil servants (Christensen 1994a).

The way the ministries recruit their civil servants has always been relatively decentralized, meaning that each ministry exercises a good deal of control over the selection process. Norway has never had a central administrative school or centralized selection process nor any sharp division between different types or classes of civil servants. This is quite different from other Western countries, and wages have over time become more equal than in other countries when compared across different levels. An exception to this pattern is the somewhat reluctant reform over the last five years that imposed a modified version of merit pay and individualized executive contracts (Laegreid 1993a).

The decentralized recruitment of civil servants in the ministries does not mean that there are no criteria for selection. In fact, all the ministries use the same types of criteria, but the content of the criteria and their combination may vary (Roeberg 1991). The most important factor for recruitment to the first job in the ministries is higher education. Most of the newcomers come directly from a completed university or graduate-school education. The social background of an individual is important for admission to higher education, but not important for the actual recruitment to the ministries (Laegreid and Olsen 1978). Approximately 80 to 90 percent of the people recruited have completed their education on a higher university level (Christensen and Egeberg 1997a).

The next step in an administrative career is to move from a position as an executive officer to become a senior executive officer. This portion of the career is protected and people normally are promoted automatically on the basis of tenure. Recruitment to different leadership positions higher in the structures is partly also based on tenure, but mainly on performance of the civil servant in his or her previous positions. The control of the selection processes follows a certain pattern: Recruitment on the lower levels is controlled by the head of a division and also is influenced somewhat by the organizations of civil servants. Selecting leaders at the middle level of an organization is controlled by the director-general, while filling the top leadership positions involves political leaders much more directly.

Changes in Demographic Structure. The composition or demographic profile of the civil servants in the Norwegian ministries has changed dramatically over time, partly a reflection of the changes in the structure of higher education but also an effect of a need for new types of expertise. The most substantial change has occurred for the lawyers. Before 1914, between 80 and 90 percent of the civil servants were jurists (Laegreid and Roness 1983, 39). In the period between the world wars this figure was between 50 and 80 percent. After 1945, the percentage of jurists has decreased from around 50–60 percent to 20 percent (Christensen and Egeberg 1997a). After World War II, a small but important group of national economists was recruited to the ministries, reflecting the emphasis the Labor Party placed on increased competence in planning, and has since then kept about 10 percent of the positions. During the previous twenty to twenty-five years the strongest increasing group in the ministries has been the social scientists, especially the political scientists, and they have today a share of the civil servants similar to the jurists. Another group increasing in this period has been the civil servants educated in business schools.

Women have traditionally been underrepresented in the central administration in Norway, but they have enjoyed a substantial increase during the previous ten to twenty years. In 1986, 26 percent of all the positions in the ministries were filled by women, but 37 percent of the executive officers were women (Christensen and Egeberg 1997a). During the previous five years four women have, for example, been appointed to the highest administrative positions in a ministry, the secretary-general.

As mentioned, many civil servants are recruited into the lowest level directly from higher education, but if they have some job experience it is normally in other parts of the public sector. We have no reason to believe, however, that civil servants are isolated from society by virtue of their backgrounds. Many of them have substantial experience in the political parties, in the principal interest organizations, or in ad hoc interest groups (Laegreid and Olsen 1978; Olsen 1983). Compared to other elites, however, the civil servants are the least socially representative. They are very heavily recruited from the higher social layers in the Oslo area; they generally have by far the highest educational level; they comprise relatively more jurists than other elite groups; and they also have a stronger connection to the public sector.

Models of Thought, Contacts, and Decision-Making Behavior. In 1975, 1986, and 1996, three large surveys concerning the central administration were conducted in Norway, providing a great deal of information on the models of thought and pattern of contacts of the civil servants (Christensen and Egeberg 1997a; Egeberg 1989a; Laegreid and Olsen 1978; Olsen 1983). One general and important finding was that some of the main results of the survey in 1975 remained valid and relevant in 1986 as well as in 1996, showing a great deal of continuity in the civil-service system of Norway.

Laegreid and Olsen (1978) based their analysis of civil-service decision-making behavior on two models: One, the representative bureaucracy model, was focused on the importance of the social background of the civil servants and how they use this indirectly in making their decisions in the ministries. The other, the responsible bureaucracy model—a Weberian-inspired model—focuses on the importance of structural factors in the ministries. This model stresses that the decision-making behavior of civil servants is based upon their structural positions and their bureaucratic careers, channeled through three mechanisms of control: Socialization is the mechanism ensuring that newcomers in the ministries internalize the main values and norms in the administrative milieu. Disciplining is the mechanism whereby civil servants are not affected by their private interests in their role enactment, because they are rewarded for doing so through a lifelong career. And in correction—a last barrier—leaders correct their subordinates or leave them with little discretion in doing their jobs. This model is also based on one element from the representative one—recruitment based on higher education.

The surveys showed that the traditional type of control, the correction mechanism, is not used extensively in the Norwegian ministries. Only one-third of the civil servants reported that they are heavily constrained by rules in their jobs, very few are corrected, and many are sending their proposals to their leaders even though they know that the leaders oppose them (Christensen 1991a). Generally, the Norwegian ministries appear to have a substantial potential for socializing and disciplining the civil servants (Laegreid and Olsen 1978). The conclusion on socialization is based on the long careers of most civil servants, and they are thereby exposed to the norms and values for a long period of time, and are also able to transfer the norms to newcomers. Socialization is also based on increased homogeneity through a mechanism of "purification": a certain percentage of the newcomers leave after some few years because they do not feel at home in the administrative milieu.

The potential for disciplining civil servants is based on many factors: First, traditionally good opportunities of promotion, based on growth and gradual recruitment. Second, relatively little horizontal recruitment, even though it is increasing over time, thereby increasing the motivation to conform. Third, the civil servants expect to remain in the ministry for a relatively long period of time and have few external job offers. In 1975, 11 percent of the civil servants had plans and offers to leave. This potentially most unstable group had increased to 22 percent in 1986, but decreased slightly in 1996 (Christensen and Egeberg 1997a).

Going more into the models of thoughts or action of the civil servants shows that the structural model, comparable to the structural-instrumental perspective outlined, explains by far the most variation. Civil servants first of all have attitudes and opinions based on their structural positions; that is, they are formally role-oriented in defense of their institutions and tasks. Further, they experience a low level of role conflict on the job and wish only small changes in their area of work. The variables that explain the most variation in their models of thought are the hierarchical level of their position, what ministry they belong to, and their main tasks or issues with which they are preoccupied. Type and level of education also explain some variations in attitudes.

The pattern of contacts of the bureaucrats are heavily role-oriented: they have broad contacts, within a well-developed specialization, according to their formal roles. These patterns vary between the ministries. There are two main clusters of contacts: One to the central, political-administrative level, encompassing contacts with the political leaders, administrative leaders in the civil servant's own sector, other ministries, and interest organizations (in public boards and committees). The other cluster is directed downwards, towards the regional and local levels, and to single persons representing themselves or their families. The first pattern is typical for the economists, and the other for the jurists.

Changing the Profile of the Ministries. Some of the changes in the ministries during the previous ten or fifteen years can be connected to a slow transformation of the ministries into "political secretariats" for the political leadership. This policy, first advocated by Prime Minister Gerhardsen in 1955, has finally been implemented during this recent period. There are many indicators of the ministries being changed into more flexible, policy-development units. First, civil servants see the Storting and cabinet as more important than before. Also, more people are spending more time on planning and policy development and less on casework. Administrative leaders also are using more time on coordination both inside and between the ministries. And there are some tendencies towards debureaucratization such as establishing more collegial bodies, more social scientists (*consequence-oriented* as opposed to *procedurally oriented*) and fewer jurists, more younger people recruited, and an increased potential of mobility. There is, however, no general, nonhierarchical trend apparent in these organizations (Christensen and Egeberg 1997a). The pattern of personal contacts is more hierarchical than ever and the collegial structures inside and between the ministries seem to be controlled by the administrative leaders.

During the last ten years the ministries have placed greater weight on the development of a personnel policy that contains more career planning, leadership training, and evaluation discussions between leaders and subordinates. This development seems to have strengthened the motivation of the civil servants (Christensen 1989). Beginning in 1991, "Management of Objectives" was also made mandatory for the central administration in 1987, but this reform seems to have problems in the implementation and has partly been transformed and differentiated and partly been played down during the past years (Christensen 1991b; Christensen and Laegreid 1996). Now there is, as in the United States, a growing interest for quality management in the central administration. In the Norwegian context this represents a return to the older emphasis on rules and procedures, of doing things "the right way."

Traditionally, interest organizations in Norway have been strongly integrated in public policy on a regular and formalized basis (Kvavik 1976; Olsen 1983). The main organizational forms have been public councils, boards, and committees, where civil servants and interest-group representatives are the main actors. This type of contact structure (called the *collegial administration*) has for many years encompassed approximately 1,000 organizational bodies, most of them permanent. During the previous ten to fifteen years these structures have been reduced by 30 to 40 percent, intending to give special interests fewer points of access to the central, public apparatus. The main interest groups, however, still have substantial influence over public policy and many other channels of contact continue to exist (Christensen and Egeberg 1997b, 185).

The Development of the Directorates: Political Control and Professional Autonomy

The relationship between ministries and directorates (agencies) in Norway has centered around two important questions: Is it right, from political, administrative, professional, or other perspectives, to delegate public authority from the ministries to the directorates? If the answer to this question is "yes," how should government organize the relationship between these bodies (Christensen 1997b)? Historically, there have been three different answers to these questions: The "Danish model" is characterized by keeping professional administration inside the ministry, but giving it a special position (Jacobsen 1960). This model emphasizes political control and integration of professional groups. The "Swedish model" systematically organizes many professional administrations outside the ministries, in so-called "independent directorates." This model stresses the need for professional autonomy and places less weight upon political control. The "Norwegian model" is something between the two other models. In this model the directorate has a dual function, both as a department in the ministry and as an independent directorate. This solution attempts to balance political control with professional autonomy. Neither of the models provides for a leader appointed on a political basis, a feature quite different from the

United States, and provides a variety of solutions to the question of how close they should be connected to the political leadership.

Historically, the first demands for a more autonomous position for the professional groups in the Norwegian central administration came during the 1820s and 1830s, especially from engineers and medical doctors (Christensen 1997a; Maurseth 1979). Approximately ten years later came the first wave of creation of independent directorates, first of all in the communications sector. But some demands in this period also resulted in integrated, Danish solutions; for example, for the health and educational administrations. The second wave of independent directorates in the 1870s spread the Swedish organizational format to other sectors. In the period from the establishment of the parliamentary principle in 1884 until World War I the number of new directorates being created stagnated. The reason for this was that the Storting before 1884 tried to undermine the cabinet by creating independent, administrative bodies, while they played on the same team afterwards.

In the period between the world wars the Norwegian model temporarily became more widespread, because of cutbacks and administrative rationalization, but this solution almost disappeared again after World War II. During the period between 1955 and 1970 a third wave of independent directorates was created, this time inspired by a doctrine of "hiving-off" professional administration from the ministries so that they could obtain more capacity for flexibility and policy development (Christensen 1997b). Since the early 1970s, Norway has experienced a more "contractive" period concerning new directorates. One reason for this was that more votes and power were held by the nonsocialist parties, and their view was that more directorates would be an expansion of the public sector. During the previous ten years the Labor Party has again obtained the upper hand, but there seems to be no clear trend in their development of the institutional policies for changing the administrative apparatus.

In 1844 there were eight directorates with twenty employees; in 1947, forty-two directorates with 3,500 employees, while the figures in 1982 were seventy-two and 9,900, respectively (Benum 1979; Laegreid and Roness 1983, 12). This means that there has been a much larger increase in the number of people working in the directorates than the ministries—namely, three times more since World War II. The structure of the directorates is, however, biased, since about two-thirds of all employees work in the ten largest directorates, and these again are dominated by the largest ones in the communication sector (telecommunications, railways, post, roads).

Historically speaking, the directorates have developed much of the same type of internal complexity as the ministries, and they are much more alike the ministries than they "should" be according to the official institutional policy; for example, in the types of tasks they perform and their discretionary freedom

(Christensen and Egeberg 1997a). The directorates recruit directly from higher education and seldom horizontally from within government. The directorates have about the same age and tenure profile as the ministries, and the civil servants' plans and possibilities for leaving are the same. Concerning decision-making premises, surprisingly enough the directorates and the ministries place equal weight on professional considerations, while the directorates might have been thought to be more professional (Christensen 1991a; Egeberg 1989b).

But directorates also deviate from the ministries in many respects: They have relatively fewer collegial bodies, hierarchical levels, and top leaders, a reflection of lower pressure on coordination; they earmark about one-third of their positions for specific professional groups; their social profile is far less biased than the ministries'—the directorates recruit more from their own policy field and the private sector, and plans and potential for leaving government are also more connected to the private sector.

The dominant mode of organizing directorates has always been the Swedish model. This solution implies that directorates have less contacts with the political-administrative leaders in the ministries, and with other ministries, than in a more integrated solution. This means also that their access to, and influence over, central decision-making processes are limited. But, on the other hand, a vertically, interorganizational solution implies that the directorates place less weight on political signals from above in their own decision-making processes, a feature that could be politically problematic, especially since the directorates have delegated public authority in many policy areas (Christensen 1997a; Egeberg 1989b). One means of modifying this effect of vertical specialization is to create "organizational duplication," meaning that the ministries have specific units working directly to the task of a directorate. Directorates also have more contacts with client groups and subordinate levels and institutions than do the ministries, and this factor also influences their decisions.

During the last decade, inspired by some market-oriented reforms, some of the directorates have been reorganized, with a looser coupling to the ministries and more autonomy economically. This trend, encompassing primarily some few directorates in the communication sector, shows that economic considerations have gained some strength compared to the political.

THE CONTEMPORARY STRUCTURE OF THE EXECUTIVE BUREAUCRACY: A COMPARISON

Both the central administrative apparatus in Norway and the federal bureaucracy in the United States have developed in societies and political systems that have grown more complex, the political-administrative issues are demanding more expertise and the political leaders are having problems of capacity. These trends have resulted in more delegation of public authority to the bureaucracy, a continuous growth in the number of people employed and an increasing specialization, both vertically and horizontally (Derthick 1990, 11). Generally, this development has increased the potential influence of bureaucracy, both as an instrument or in its own right, but also has made it more open and interesting for other political-administrative actors.

Norway and the United States are also similar concerning the demography of the civil service. Even though the United States has a Office of Personnel Management (formerly the Civil Service Commission), recruiting to a classified service, the United States like Norway has a decentralized, institutionally based merit system of recruitment of civil servants (Martens 1979, 238). The Civil Service Reform Act of 1978 created the Senior Executive Service, an attempt to obtain more competent and professional top leaders, and has its counterpart in Norway in the creation of merit pay and contracts for the top administrative leaders (Laegreid 1993a, b; Schuman and Olufs III 1988, 154). The civil servants in both countries have very often lifelong careers inside the same sector or institution, not moving around too much in the administrative apparatus. Further, the more specialized parts of the bureaucracy appear to recruit people with relevant education and background from the private sector (Peters 1989, 81). The modes of thought of the civil servants seem mainly to be influenced by their own profession, program, tasks, and department/agency (Laegreid and Olsen 1978).

In the federal bureaucracy in the United States, recruitment has tended to be of specialists in the policy area they will administer, although there is some increase in generalists managers (Nathan 1983, 86; Page 1992, 45–48). The Norwegian central administration is characterized by a former dominance but sharply decreasing representation of jurists, a growing proportion of economists, and especially social scientists, and a large share of people educated in the natural sciences in the directorates (Christensen and Egeberg 1997a; Olsen 1983, 125).

The federal bureaucracy in the United States is, however, much more multistructural and fragmented than that in Norway, characterized by a complicated pattern of conflicts and cooperation, both internally and externally. This is, of course, a reflection of a larger, more divided society and political system, a Constitution that is characterized by a separation of powers, and a definition of the federal bureaucracy that is diffuse, not providing it any solid and legitimate basis of participation in the governance process (Rohr 1987, 114).

Based on the Constitution and its own development in different historical phases, the federal bureaucracy today is a mixture of many different considerations (Morrow 1987, 162–71). At first, the control of the executive bureaucracy was mainly a task for Congress, leaving relatively little influence for the president (Sundquist 1987, 263–64). Later, the federal bureaucracy began to develop, based on both a division and later integration (Jacksonian era) between politics and administration, and a doctrine of *executive prerogative* as an instrument for the president was created. The president was expected to organize his own executive branch—appoint, assign, and motivate, and thereby penetrate the administrative decision-making process (Nathan 1983, 82).

In this century the president has gradually developed, through delegation from Congress, as a sort of general manager for the federal bureaucracy, but Congress has retained considerable authority in a role as some kind of board of directors (Sundquist 1987, 266), and has been engaging in increased levels of micromanagement. Congress has never fully supported the theory of presidential responsibility over federal bureaucracy, trying to supervise the bureaucracy closely, and arguing that the delegation of lawmaking and spending power could be withdrawn or overruled, thereby saying that it acts constitutionally (Derthick 1990, 17).

Taken together, this leaves the federal bureaucracy subject to cross-pressures; it is more a battlefield between the president and Congress, and also the judiciary and the interest groups. This is intended to further fragment and undermine presidential control of the bureaucracy. This is seen by many scholars more as a problem of political leadership than as a problem generated by the executive branch (Van Riper 1987a, 31). The federal bureaucracy has also passed through different phases emphasizing neutrality and professionalism, and then clientelism, seen by some observers as closely connected and by others as incompatible (Morrow 1987, 169–72).

In Norway, the central bureaucracy is much more an integral part of the executive, a role that is logical in relation to the parliamentary principle and which also provides the administrative apparatus with higher prestige than in the United States. In the United States, the federal bureaucracy must serve a number of different masters such as the president and Congress, and enjoys considerably lower trust and prestige, both inside and outside the political leadership. Certainly, at the present time, the bureaucracy is a convenient "whipping boy" for both the president and Congress.

The structure and demography of the central bureaucracy in Norway and the United States seem to have some other differences also: First, in the United States the president is directly responsible for the activities of every single unit in the federal bureaucracy. In his day-to-day work the president can, of course, not monitor all the activities, and must delegate and use his political appointees. But if anything problematic happens, he is to blame, and this is one main reason why the president therefore can use much power on a short-term basis to change institutional conditions. The formal lines of control are, however, more complicated than this, since Congress—more specifically, committee and subcommittee many times act "as if" they were in command of organizations. Based on their legislative powers and the power of the purse, they can sometimes closely follow-up the activities of individual units in the federal bureaucracy.

In Norway, the lines of authority are much clearer and more hierarchical. The central administration is an integral part of the executive, and the legislature is very seldom involved in its activities. Norway has a system with ministerial responsibility that implies that direct control and criticism of the bureaucracy seldom occur. Very often the authority is delegated to lower level of the ministries and to the executive agencies, the directorates, in fact implying that many decisions are taken at the agencies' own discretion (Christensen 1997b). An important background for such a system, and also a factor strengthened by such a structure, is the trust in the professional quality and expertise of the civil servants (Laegreid and Olsen 1978). And, of course, also that they are reasonably sensitive to the political signals in their discretionary behavior (Christensen 1991a).

Second, the structure of the federal bureaucracy in the United States is much more heterogeneous than in Norway, implying that there are many more different types of bodies (Martens 1979, 224). Seidman and Gilmour (1986, 22) provide a long list that can illustrate this very well: executive departments, independent agencies, assorted commissions, boards, councils, authorities, wholly owned corporations, mixed-ownership corporations, "captive" corporations, institutes, government-sponsored enterprises, foundations, establishments, conferences, intergovernmental bodies, compact agencies, and a wide variety of interagency and advisory committees. This seems to create problems of control by the president, both possibilities and disadvantages for Congress, but many points of contacts and potential for influence by the interest groups (i.e., it is widely open to a pluralist pressure) (Morrow 1987, 162).

In Norway, there are mainly two types of bodies in the central administration, the ministries and the directorates, even though the last decade has provided some organizational forms that are more similar to some of the executive agencies in the United States (Christensen 1991a). Norwegian administration is a much more closed and homogeneous structure. Formal access of outside actors, mainly interest groups, is more regularized and formalized through the so-called "collegial administration," corporate bodies consisting mainly of civil servants and representatives from the interest groups (Christensen and Egeberg 1979; Egeberg 1981; Kvavik 1976; Olsen 1983). These bodies are, first of all, advisory bodies and participate in the preparation of lawmaking but seldom have any decision-making authority.

Third, in the United States the political appointment system—namely, the president appoints agency leaders and even SES members that are his representatives—has traditionally created a much more politicized central bureaucracy than in Norway. These appointments have meant that the career system of the civil servants has been more politicized (despite the doctrine of neutral competence), unstable and ambiguous (Van Riper 1987b, 31). The president has under this system to seek advice and consent from the Senate for appointments, a procedure that makes him and the federal bureaucracy (meaning here *appointees* and not the career service) more vulnerable (Martens 1979, 237).

In Norway, there is no formal political-appointment system such as in the United States, even though ministers and their few political helpers do leave office when the cabinet is changed. But most of the administrative bodies and their civil servants, located in the directorates, have no day-to-day political leaders and few such connections, and are headed by career civil servants, or in some cases by former politicians who have by then adopted a clearly administrative role (Christensen 1997a). The political significance of appointments is also reflected in a different use of staff units in Norway and the United States. In the United States, the staff connected to the secretary of the department participates much more in the regular control of the bureaucracy than do similar staffs in Norway (Martens 1979, 212). In Norway, the use of staff units is traditionally much more limited, and the staff has first of all advisory and supporting tasks; Norway has a tradition with more line organizations operating with limited staff overhead.

Instruments in the Executive Control of Bureaucracy

To control the bureaucracy, politicians have primarily two types of mechanisms: on the one hand, politicians can impose external means of control; on the other, they can rely on the more internal checks, resulting from the discretionary behavior that have built-in considerations in favor of political signals (Riley 1987). With growing complexity of public issues and increasing work load on the central political leadership in many countries, the leaders face serious problems of control capacity. One central way to try to solve this problem is to combine more general laws with delegation of authority to the central bureaucracy, thereby giving the bureaucracy more freedom or discretion. This makes the mechanism of internal control in the bureaucratic roles more important.

The distinction between internal and external control seems to emphasize precisely a principal difference between the central bureaucracies of Norway and the United States. In Norway, the external means of controlling the bureaucracy are relatively few and come mainly from a single source, the executive political leadership. But internal bureaucratic controls appear to function very well, since the civil servants have few problems in balancing political and other considerations in their roles (Christensen 1991a; Laegreid and Olsen 1978). In the United States, on the other hand, the instruments of external control are numerous, coming from different sources and are often inconsistent (Riley 1987). This makes the conditions for internal control worse, thereby giving the civil servants more problems in their role enactment and leading to a more active political role (Aberbach et al. 1981, 94–101).

One central premise in the political-administrative system of the United States is that presidents for a long period of time have mistrusted the federal bureaucracy. This is, first of all, a reflection of the political culture, where the public bureaucracy in general has a rather low status and little prestige. But it is also believed to be an effect of the separation-of-powers system and the growing heterogeneity inside the executive power.

Many studies emphasize that the federal bureaucracy, through the increase of its discretionary influence, has undermined the political control of the president in at least two ways: First, the agencies have participated in alliances or "iron triangles" with subcommittees in Congress and client groups (Gormley 1989, 6–7). Second, agencies (especially regulatory agencies) are said to be captured (i.e., to have become prisoners of strong client groups originally meant to be regulated) and this has lessened presidential authority. This development has been furthered by both a more activist, well-staffed, and competent Congress, and by demands for more popular participation in government.

Gormley (1989) lists an impressive number of means of control over bureaucracy. He divides them into three categories: prayers or catalytic control (i.e., the expression of discontent and hope of bureaucratic change); hortatory control (i.e., more pressure but also some incentives); and muscles or coercive control (i.e., control through hierarchy and command), a type believed to be of increasing importance during the last two decades. Different presidents have tried to use various combination of these types of control: First, they have tried to use administrative reorganization to get control (Gormley 1989, 117–20). The effects of this appear to be most evident when the bureaucratic discretion is at its highest and the involvement of Congress at its lowest. But presidents often become less interested in reorganization after some time in office, realizing that they can win few political victories with this instrument (March and Olsen 1983).

The reorganization processes have both encompassed the restructuring of executive departments and agencies, but also structures of coordination in the political leadership such as counter-bureaucracies in the Executive Office of the President, interagency task forces, and different cabinet councils. A variant of administrative reorganization has been an increasing politicization of personnel policy and reform. The president has also appointed inspector generals, a rapidly growing structure of units of oversight, evaluation, investigation, auditing, and control in the agencies. A third set of means has been executive orders from the president: The direct control over the agencies through rules that have the same effects as formal law but not explicitly the consent of Congress. Executive orders are often used by the president in situations of conflict under divided government.

The increased representation of interest groups in bureaucracy was originally a result of the Democratic-inspired reforms starting twenty years ago in the United States. These reforms open up a wide variety of possibilities of influence:

Increased access of information, sunshine laws implying participation in collegial bodies, public hearings, public support for the intervention of interest groups given certain conditions, surrogate representation through Ombudsmen and proxy advocates, and so on (Gormley 1989, 62). The danger is, however, that they simply become another avenue through which capture can take place.

In Norway, the status of the central bureaucracy and its civil servants is very high, both among politicians and the general public. The bureaucracy and its expertise are seen as stable elements in the democracy and as safeguarding the quality of public decisions (Olsen 1983). The traditional Weberian means of bureaucratic control are very seldom in use, and the mechanisms of control are first of all that the civil servants are internalizing the main values and norms in the administrative milieu, and disciplined through the gradual incentives of lifelong careers (Laegreid and Olsen 1978, 119–27; Olsen 1983, 126–30).

In such a system of trust and peaceful coexistence between the politicians and the bureaucrats, one important way of communicating from the top-down is to give political signals. These signals seems to be received and used by the civil servants in a unproblematic way (Christensen 1991a). This illustrates an important scholarly point made many years ago, namely, that the Norwegian bureaucrats are good at balancing different important elements in their bureaucratic roles (Jacobsen 1960). This means, first of all, the balance between political loyalty and neutrality, and between the professional autonomy and professional subordination. More generally, the civil servants also have to attend to institutional values and norms ("the voices of the past"), citizens rights and areas of no public regulation, affected parties or clients, and so on (Egeberg 1984).

One of the main points made by Jacobsen (1960) is that the somewhat ambiguous mixture of different considerations in the bureaucratic role is good for the bureaucracy, the politicians, and the democracy. A more precise ranking and weighing of the considerations would eventually generate much more political and administrative conflicts. The use of political signals as a means of control and direction has generally more effect in the ministries, the apparatus closest to the political leaders (Egeberg 1989a, 100–103). The civil servants in the directorates are far more removed from the political signals and put less weight upon them (Christensen 1991a, b).

The relationship between the politicians and bureaucrats in the central administration in Norway is to an increasing degree based on laws of a general character and characterized by delegation of authority. This is reflected in the fact that two-thirds of civil-servant respondents in surveys say that they have some or a great deal of discretion in their daily work, and that there is growth in the groups' having a means–end expertise (Christensen and Egeberg 1997a). This does not mean that political control is undermined, because more civil servants than before are telling that they are engaged in hierarchical coordination activities, and the bureaucrats are reporting a clear hierarchical pattern of contacts inside the bureaucracy and in relation to the task environment.

The Norwegian political-administrative elites are, like those of the United States, continuously using reorganization as an instrument to further political goals (Christensen 1987, 1997a; Egeberg 1984, 1987, 1989b; Olsen 1988b). They have attempted to create organizational attention through establishing new organizational units or changing their decision-making behavior through merging or dividing units and so on. Civil servants very seldom oppose such processes of organizational change. The more politically important a reorganization is (such as establishing a ministry or making a major change in the structure between a ministry and a directorate), the more centralized and closed is the pattern of participation (Christensen 1987; Egeberg 1984). Most of the processes of reorganization are headed, however, by administrative leaders that are implementing the changes on behalf of politicians.

The importance of financial control in the Norwegian central administration is relatively high, especially during periods of contraction and decline. This is, however, mainly a general and on-going type of control exercised by the Ministry of Finance. The ministries are mainly allocated the resources they ask for, minus some cuts that are applied across the board to all units, and they also have a reasonable degree of freedom in using the resources. During certain periods, the political leaders are earmarking resources to certain programs, but this is normally a very small proportion of the total budget.

Summary

The contemporary administrative systems of Norway and the United States represent the cumulation of their historical development. In both countries there has been increasing complexity, although both countries have inherently rather complex systems. This is because they have highly differentiated administrative systems with numerous boards and agencies enjoying substantial autonomy from their departments. Recent attempts at reform in the two systems, and especially within the United States, have attempted to streamline government and reduce some of that complexity. These changes are perhaps more needed in the United States, given the redundancy built into the system. Even with the changes in progress, both administrative systems are extremely complex and somewhat redundant at times.

A second general point about the development of public administration in the two countries is that there has been an increasing level of politicization (Light 1996). Again, the United States is the more extreme case. It has had a more politicized administrative system than Norway for most of its history, and the changes during the last fifteen years have tended to reinstitute even more of a "spoils system" than existed previously. Politicization has been extended to civil-service positions that had previously been protected from political appointments and interference.

Finally, there are significant differences between the two administrative systems, just as there are important similarities. Norwegian public administration is a respected, elite body that is extremely influential in formulating as well as implementing public policy. The civil service in the United States is not a respected institution, but despite that, it also has a significant impact on the policies adopted by the federal government. Both government systems provide organizations a great deal of autonomy, but that freedom is legally based in Norway, but is based largely on political relationships with Congress and its committees in the United States. Every effort is now being made to rein in some of the power of the bureaucracy, but it is not clear if government can do its job, even become smaller, without the active involvement of the civil service.

NOTES

1. This model was terminated, after numerous earlier attempts, in 1995.

2. Under the reorganization, the War Department became the Department of the Army.

3. The caretaker presidency of Gerald Ford intervened, of course, but had little appreciable impact on public administration. The one major act of relevance here was the Congressional Budget and Impoundment Act of 1974.

4. The specialized, technical parts of the service were exempted from political appointment in order to ensure their technical competence.

5. These organizations have since been terminated as a result of the Federal Pay Act of 1990.

6. Perhaps the major event of relevance of the Bush administration was the Pay Act of 1990, which broke the connection between civil-service and congressional pay.

7. The Gore Report itself was quite clear, however, in saying that the problems it detected were not with managers themselves but rather in the multiple rules within which they had to function.

CHAPTER SIX

Cultural Variables and Governance in Norway and the United States

IN CHAPTERS 2 THROUGH 5 the formal structure, both inter- and intraorganizational, of the central political and administrative systems in Norway and the United States has been analyzed, and its effects on decision-making behavior and governance discussed. One main conclusion was that the United States is generally a structurally very fragmented and heterogeneous political-administrative system, while Norway is homogeneous, both features that partly is the result of a conscious organizational design. On a general level, we have also touched upon cultural norms and values in the two systems, because we have discussed the type of culture that can develop inside such structural arrangements, namely, that the United States seems to develop a inter- and intrainstitutional culture that is characterized by heterogeneity and conflict, while there is more collaboration and less conflict in the Norwegian system. This chapter, fully concentrating on the cultural variables, will on a more broad basis discuss how cultural norms and values can be nested, how they can constrain and influence structural factors, and how cultural and structural factors can work together and constrain and channel decision-making behavior and influence the capacity of governance.

Our point of departure is that as well as being influenced by the formal structures of their political systems, performance of Norway and the United States in governance is also influenced by cultural factors. There is a dangerous tendency in comparative political analysis to attribute any observed differences among systems not readily attributable to more directly quantifiable factors to differences in "political culture" (Elkins and Simeon 1979; Street 1994). This cultural reductionism often reflects the failure either to specify or to measure political culture adequately, with the consequent need (and ability) to include it in analyses only as a residual variable. If the other hypothesized explanatory variables do not perform as anticipated, then it often is assumed that culture *must* be the cause of observed differences, and that becomes the end of the search for an explanation.

We will assign importance to cultural factors in understanding the differences between the United States and Norway, but we also attempt to specify the dimensions of culture that we consider important and provide some measures of the concept and its dimensions. Those dimensions can help disaggregate an otherwise excessively broad concept, and thereby help to understand just how cultural values and norms *do* impact the performance of a government. Further, having some predetermined dimensions along which to compare cultures can help prevent a cultural argument from degenerating into national stereotyping. Inevitably, any attempt to characterize national values will have some elements of stereotyping, even when executed by a member of that nation (Seigfried 1940), but we will attempt to minimize those problems by a careful attention to the literature on political and administrative culture, attempts at using the same dimensions in both cases, and sensitivity to the possibility of overstatement of what are generally subtle differences, especially among industrialized countries.

As we discuss the influence of culture on administration and governance, we will differentiate the concept along several levels of generality. Although discussed separately, these levels should be conceptualized as nested, so that those at a higher level of generality exert some influence over, and contextualize, those existing at lower levels of generality. First, we will look as some features of general social and political culture that influence the capacity of the governing system to make and implement its decisions, as well as perhaps also influencing the actual content of those decisions. The most general of these are the cleavage structure and factors connected to Douglas's cultural theory (1986, 1990). On a less general level are factors related to political and administrative culture, like attitudes towards legitimacy, attitudes towards law, judicial expertise and due processes, and trust in institutions and individual actors. While these "ideational factors" in social life sometimes appear amorphous (Schulman 1991), they also are pervasive so that attempting to understand political and administrative differences without at least some acquaintance with those public ideas may be a very ill-fated enterprise. The ideas and concepts about "good government" that both citizens and decision-makers carry around in their heads ultimately may be as influential as any more tangible factors such as structure.

We will then discuss some more specific, cultural variables. First, we will focus on cultural traditions connected to networks of public organizations, intrainstitutional culture or culture related to policy fields and professions. Second, specific cultural elements surrounding public management will be discussed. Lastly, factors related to individual roles will be discussed: consistency in role elements, exposure and definition of accountability, and discretion. One reason for including such specific variables is that there is a tendency—for example, in the "state literature" (Skocpol 1979; but see Evans 1995)—to analyze governments as undifferentiated wholes, whereas in reality administrative systems tend to be highly differentiated internally, both between institutions, inside institutions, and between roles (Muller 1985). The structuralist perspective being employed in this analysis requires that we think about more differentiated administrative structures and their influence on policies. We are arguing that structures tend to embody values that their members, in turn, utilize to make decisions about public policy and delivery of services to clients. This is one of the fundamental perspectives of institutional analysis, and central for understanding policy-making. As is true for all levels of culture discussed, organizational culture is difficult to document (but see Hofstede 1991) and measure, but it nonetheless is an important element in understanding the outcomes of governance in complex democratic governments.

SYSTEMIC LEVEL OF CULTURE

As noted, we will begin by identifying some cultural variables that operate at the systemic level in all countries. These are factors that exert a pervasive, if often diffuse, effect over policy and administration, and over the behavior of politicians generally. Because these values are largely environmental, these cultural factors are often difficult to identify and are also subject to stereotyping and overstatement. The creation of national stereotypes—phlegmatic Englishmen, excitable Italians, somber Swedes—is but one of the more undesirable consequences of this tradition of political analysis. Despite those potential pitfalls, we believe that it is important to discuss these ideational factors for Norway and the United States and attempt to show their impact on governance in the two systems. In the usual litany of causal factors—institutions, interests, and ideas—it is important to include ideas—including cultural values—just as we already have institutions. This explanation will run the danger of creating stereotypes, but since we are doing the stereotyping of ourselves it is perhaps somewhat more permissible. Further, attempting to understand political behavior without including values may be as unacceptable and unwise as relying too heavily on values and culture.

Social and Economic Cleavages

All societies have some social and economic cleavages with political relevance. These cleavages may be based on economics (wealth, occupation, etc.), geography (rural versus urban, regionalism), or ethnic characteristics (race, language, religion, etc.). Further, these divisive factors often reinforce themselves, as when an ethnic group tends to be concentrated in one region of a country and to be poorer on average than the population as a whole. That reinforcement tends to exacerbate divisions and make the task of governance all the more difficult. Whether they reinforce each other, as in the United States, or not, the basic impact of these cleavages is to divide segments of the society from one another. The intensity of those divisions may be quite low, as in Norway, so that national integration is not threatened, but the divisions do create a basis for political activity and policy. The characteristics of these socioeconomic cleavages will have at least three significant effects for this study of policy-making and administration.

First, the number, type, and intensity of cleavages will influence the political "load" that a system must bear and process. In a society such as the United States's, with a number of deep and often reinforcing social cleavages, the government simply is presented with more problems to solve than it might in a less complex social system. The government of Norway, in contrast, will have somewhat fewer policy problems coming from the social system, especially as urbanization and general social change have gradually weakened the "counter-culture" (nonalcohol, new-Norwegian, and religious) that has existed outside the major cities. Still, it certainly does have socioeconomic cleavages that have provoked a significant amount of scholarly discussion (Rokkan 1970) as well as real political discussions. Further, the types of socioeconomic cleavages characteristic of the United States may be more difficult for any government to cope with than would be true for those encountered in Norway. The cleavages of language, and especially race, are extremely difficult to process politically once they become highly politicized. This is in large part because they cannot be bargained away as easily with money as can the socioeconomic cleavages that tend to dominate in Norwegian politics.

Most of the cleavages of important political relevance in Norway have an economic base. Even some of the cleavages that appear based on regional interests have at their roots a desire to preserve a rural and agricultural way of life in the face of threats from urbanization and industrialization (Lafferty 1981). Further, these economic cleavages (especially when conceptualized as based on occupation and industrial sector rather than class) align closely with other major cleavage dimensions in the society such as religiosity and language. These noneconomic cleavages tend to be declining in importance and, that being the case, the cleavages system is more likely to be bargainable, with public money generally being the medium of the negotiation. The Norwegian government has a clear record of attempting to maintain the commitment and equanimity of potentially alienated segments of the population by using large transfers of funds to locales dependent upon the primary economy as their instrument of preserving the political peace at the same time as preserving more traditional ways of life. In contrast, many deep cleavages encountered in the United States are based on race and language as well as to some extent on religion. Many of these cleavages also contain an economic element, but at heart are issues of identity and cultural recognition that may go well beyond the capacity of money alone to paper over. As a consequence, any government in the United States will, *ceteris paribus*, have to do more to manage its cleavages effectively than will a government in Norway (Lijphart 1991). The civil rights movement of the 1960s and 1970s did a great deal to produce greater de jure equality and tolerance among the races in the United States, but has been somewhat less successful in creating those positive outcomes de facto. Further, demands from African Americans and other ethnic groups for a multicultural society can easily be interpreted as demands to *widen* the gulfs already existing between racial groups, rather than narrow them (Hacker 1992; Schlesinger 1992).

Further, religion is an increasingly important source of cleavage in American political life (*Public Perspective* 1994), despite the attempts of the founding fathers to create a "wall of separation" between church and state.[1] At least one of the most important political issues (abortion) of the 1980s and 1990s has a very strong religious and moral component and the religious right has been willing to go to great lengths to impose its views of morality on the rest of the country. Further, the religious right[2] in the United States has become increasingly mobilized politically and has been able to take over the Republican party in some states and localities, and perhaps even the national party. This activism has tended to place religion and its role in politics much more in the center of political life in the United States than in the past, and may be opening some deeper cleavage in a society that has long prided itself on its religious toleration. The religious right has been intolerant and appears to be promoting intolerance among other segments of the population. Norway is saved from the worst potential of this cleavage by virtue of having an established state church and a traditional of tolerance for dissenters.

A second factor about cleavage of relevance to governance is that the factors that divide members of the mass public tend to be reflected in the political and administrative systems themselves. Further, there is an on-going, dynamic relationship between cultural and structural factors in governance. Cultural factors (especially socioeconomic cleavages) tend to engender socioeconomic structures, and structural elements of the system are sometimes introduced into political behavior by cultural change. The socioeconomic pluralism of the United States now has been reflected in a series of programs and institutions attempting to "solve" the socioeconomic problems of the society; for example, the Equal Employment Opportunities Commission. These structures then further divide an already divided decision-making system and add to the already substantial internal complexity of decision-making. In Norway, there is a pattern of "institutional pluralism" that also reflects the multiplicity of societal interests in the formal structures of government (Egeberg 1997), but which can be managed within the agreements on bargaining and accommodation.

An example of the link of structure and culture can be seen easily in the United States. The U.S. Congress has been increasingly divided and subdivided to represent all the diverse interests in society, with the consequence being internal gridlock, as well as gridlock with other institutions. The Republican victory in 1994 had as one of its first consequences a significant reduction in the number of committees and subcommittees, and especially the subsidized caucuses, in Congress, with an explicit attempt on the part of the new majority to minimize the impact of interest groups and to create a more homogenous political culture within the United States. The Republican leadership had, of course, a very clear image of what that common culture should be, but were explicitly manipulating structures to attempt to facilitate its reassertion.

This increased social divisiveness is all the more evident given that the manner in which these deep social cleavages are handled often is not that of compromise and bargaining but rather that of confrontation. The most obvious example of this lack of compromise is the role of the courts in the resolution (or management) of racial conflicts such as segregation, and of religious questions such as abortion. The Anglo-American legal system is based on adversarial principles rather than mediation or fact finding (Kelman 1982). The most fundamental assumption is that the best decisions are reached by contending sides marshalling their evidence as well as possible and one winning and the other losing. This system provides relatively little room for compromise and bargaining, with the important exception of plea-bargaining used to clear up many cases before trial.

In contrast, the political-administrative system of Norway does not have to deal with many issues that are perceived to be so deeply divisive, and therefore can afford to depend upon a more streamlined and homogenous administrative order.[3] Certainly there are some religious differences within the Norwegian population, with some parts of the country displaying a good deal more religiosity than others, but these rarely manifest themselves in the vocal and even violent conflicts found in the United States over these issues. Similarly, there are some language disputes which, although very important to the people involved, are more moderate than those found in many other countries. Most of the issues with which the Norwegian government must contend are bargainable and divisible.

Finally, the cleavage structure of a country can affect its capacity to implement policies. This effect may occur through a variety of means. First, the more fragmented the culture, the more there will be demands to permit groups, especially minority groups, to have some control over the implementation of programs in their own communities. If policy uniformity is an important goal for government, then this form of participation is likely to be an impediment to attaining that goal. Likewise, the actual process of implementation itself is likely to be compromised by differences in cultural values if controlled by groups, depending upon how strong and pervasive the dominant culture is (Krasner 1988). Policies and actions will have different meanings in different cultural groups, so that a policy that it perfectly acceptable in one setting (abortion?) is much less acceptable in others. Not only will policies be unacceptable in some cultural settings, but the individuals who are charged with administering programs may be themselves unacceptable to the community that is the target population for the program. The case of white social workers in African-American communities is an obvious example, tensions between Hispanic and African-American groups over educational and housing programs is another.

As might be expected, the homogenous Norwegian culture is likely to present fewer barriers to implementation than is the more pluralistic American culture. But even Norway experiences conflicts over allegations of cultural-insensitive public programs for immigrant groups. Indeed, two of the impediments to governance mentioned above as examples—policy attitudes and racial differences—have obvious relevance for cultural politics in the United States. Further, the increasing cultural pluralism in the United States resulting from continuing immigration is likely only to exacerbate this problem for the foreseeable future. Language itself, as the medium of most implementation, is also coming under strain as a smaller proportion of the American population speaks English as a first language.[4] An important cultural change—the decline of the melting-pot idea in favor of multiculturalism (at least among some elite groups)—will mean that the plurality of the social and cultural systems is likely to be even more disruptive in the future. Immigration into Norway is also increasing, but it is at a very low rate as compared to the large numbers of people entering the United States legally and illegally each year. Already the pluralism in Norwegian society is reflected in the enhanced activities of several ministries responsible for controlling immigration as well as assisting legal immigrants.

Group and Grid: A Theory of Culture

Another way in which to approach the general cultural values that influence political and administrative life is to apply the now famous "group-grid" theory of anthropologist Mary Douglas (1986, 1990). The basic idea in her theory is that societies can be characterized by those two dimensions and that the mixture of the two attributes will influence how individuals interact with one another and with other social actors. One of the most significant features of this theory of culture is that it tends to place individuals and individual behavior at the center of the explanation, rather than depending on sweeping characterizations about collective values and norms, based largely upon assumptions about how people interpret those values.

"*Group*" is the first attribute used to characterize cultures in the Douglas approach. By group, she means the extent to which individuals believe that they belong to particular social groups; this dimension then would not be very much different from the discussion of social cleavages immediately above. In this instance, however, the groups in question may be social or they may be groups (clients, taxpayers, or civil servants) formed as a result of state action. This approach therefore can accommodate social differentiations that extend beyond the usual socioeconomic roots of politics to cover the range of different status configurations in contemporary society.

As we will point out below, one important characteristic along this dimension will be the number of different groups to which one individual may believe that he or she belongs. There will be important political differences if an individual believes that there is only one relevant group or whether the individual can shift memberships readily among a number of different groups. This ability to shift allegiances would be especially important if individuals are able to participate in groups with a variety of different types of individuals. Multiple and overlapping group memberships can reduce the sense of "us versus them" that often exacerbates social tensions, and that also reduce the capacity for governance.

The "grid" dimension of culture refers to the perception of the appropriate extent and variety of rules in society, a dimension that is closer to the activities of public institutions. In other words, do individuals believe that social behavior should be (or is) determined largely by rules (formal or informal), or is there greater latitude for individuals to make their own decisions about what behavior is appropriate? Also, does the individual perceive that there are uniform rules, or is there a variety of different rules regulating behavior by different people and/or in different settings? The more decisive rules are perceived to be, and the more differentiated sets of rules that are perceived to exist, the higher the individual will be said to be located on the grid dimension of culture.

The interaction of these two dimension is argued to produce four basic cultural patterns: Individuals (and by extension whole cultures) that are high on both dimensions are said to be "hierarchical." Individuals perceive themselves as belonging to a group, and that group has its interactions with other groups determined or shaped by a rather elaborate set of rules and norms. Individuals who rank "low" on both dimensions are described as "individualists." They tend to regard themselves as atomistic actors whose behavior is to be guided primarily by internalized norms and values rather than by group membership or collective norms. Individualists, therefore, would appear to correspond to the familiar model of "economic man" as utility maximizers relatively unconstrained by collective values.

The mixtures of high and low rankings on these two dimensions produce two other interesting cultural patterns: High group and low grid values produce an "egalitarian" cultural pattern. In this pattern there is a strong sense of group membership, but few rules that are perceived to govern interactions within or between groups. This absence of effective and enforceable societal rules, in turn, appears to require consensual decision-making within the group. Thus, after a group has made its own decision, there will be limits on the possibilities of compromise with other groups, and that may make the role of the groups in political life also more difficult.[5] If entire political cultures are egalitarian, then there may be real governance problems in generating compromise across parts of the society.

Douglas and her followers describe individuals (or perhaps societies) with high-grid and low-group values as "fatalists." As with the individualists, these individuals do not perceive themselves as belonging to any group, but yet they feel that their decisions are governed by a host of rules not of their own choosing. Thus, they believe that they have little control over their own lives. In some instances, such as prisoners or military recruits, they may be correct. This pattern resembles Etzioni's (1975) idea of "alienative compliance" in organizations, where the individuals comply but only because of the threat of retaliation if they do not, rather than from any acceptance of the values of the institution imposing the rules. While egalitarian cultures present one form of governance problem, fatalist cultures would present another, given the absence of any real commitment to the values of the political system.

If we look at the individual level of analysis, then each person may display some or all of these four types of cultural patterns, usually at different times and over different issues. Still, we would probably be able to identify individuals as being more or less of one type or another. Likewise, almost any society of any size at all would have at least some individuals who could be classified as all of these types. Again, however, we would usually be able to typify the society as being dominated by one type or another of cultural perspective. Even if we are not able to classify entire societies, we could probably identify significant subcultures or identify certain circumstances in which the dominant pattern was one or another of the cultural groupings. We have already noted that we are sensitive to the dangers of stereotyping any time that cultural explanations are used and cultural descriptions are advanced. Still, there is substantial utility in engaging in this exercise for the two countries in which we are interested, and pointing out how these attributes influence political behavior.

The United States. The dominant cultural pattern in the United States would appear to be individualism. The stereotype of American culture certainly is that of "rugged individualism"; the location of the culture in the group-grid matrix appears to confirm that stereotype. In the first place, one of the dominant myths of American society has been the "melting pot" in which all of the various immigrant groups that came to those shores were transformed into just "Americans." Certainly ethnic and cultural allegiances from the "old country" may persist, and persist for generations, but the dominant feature was the common allegiance to the national entity, and with that a political and cultural commitment to individual freedoms. The pluralistic practice of politics has tended to reduce the importance of group ties, although certainly not to eliminate them entirely.

Certainly the economic culture of the United States is more individualistic than is that of most other industrialized democracies, to the point of permitting only minimal levels of public intervention into social and economic life. The failure of Congress to pass health-care reform in 1994, and the debate that surrounded that failure, are indicative of the lack of commitment to social intervention, even in such crucial areas of personal life. The electoral success of the right-wing of the Republican party in the 1994 Congressional elections is another clear indication of the power of individualistic thinking in the United States. Further, more Americans appear to be "bowling alone" (Putnam 1995), continuing to participate in activities but on an individual as opposed to group basis.

Although it appears appropriate to argue that the United States does have an individualistic culture, there are several challenges and caveats to that characterization. One of the more important contemporary challenges is multiculturalism. The claim is being advanced more frequently that not only is the melting pot little more than a convenient myth for American society, it would be undesirable even if true. This argument would have the United States to be a multicultural society in which there are strong group allegiances that persist in the face of strong homogenizing pressures. As noted, this argument is both empirical and normative, saying that the society has always been more group based than is thought (especially when considered along racial lines), and that it is important for some segments of society, especially ethnic minorities, that the society and culture remains divided in that way.

Another challenge to the argument on behalf of American individualism is the strong associational life existing in that society. Both the United States and Norway are characterized by a large number of voluntary organizations that play significant roles in social and economic life. Can a culture characterized by individualism produce a society populated by "joiners" and volunteers? The evidence appears to be that it can, although the cultural meaning of group membership may be different than in other cultural settings. Other than religious organizations, the groups to which Americans tend to belong are not those that hold a large sway over the lives of their members. For example, although group membership in the United States is high by international standards, membership in labor unions is comparatively low. Whether restrained by laws that are not particularly favorable to unions or by their culture, the working public does not depend upon group membership to fight their economic fights to the extent common in other industrialized countries. Further, American memberships tend to be more locally oriented, with many people belonging to national organizations only as a by-product of having joined a local or state organization. Although still, this membership pattern does not contribute to as strong a political commitment as might be true in Norway or elsewhere in Europe.

A final apparent anomaly for the United States is the importance of the Constitution in the political and governmental life of the country. If individualism is in part defined by the relative paucity of rules, then it does not seem logical to have so much political discourse in the United States governed by a formal statement of the rules of governing. Can this apparent anomaly be resolved? One way would be to say that these cultural characterizations are inevitably tendencies rather than perfect predictions of behavior so that some exceptions are almost inevitable. Further, any group or society will need some sort of rules to guide behavior, and the Constitution does that for the United States.

Another answer, and less of an overt "cop out," would be to argue that the Constitution is itself an individualistic document. Certainly, scholars working from an economic-determinist perspective have argued that the constitution was designed to ensure the freedom of individuals, especially in their economic lives, so that the major restraints in the document were on government rather than on individual behavior (Beard 1935). The importance of constraints on government can be seen most clearly in the Bill of Rights (the first ten amendments), but also shines through in some parts of the body of the document as well. Indeed, the elaborate set of checks and balances built into the policy process prevents American government from acting swiftly and decisively to regulate individual action (Fiorina 1992). Those restraints on action have been exacerbated by "divided government," with the presidency and Congress controlled by different parties for much of the last fifty years.

One of the strongest indicators of the individualistic bent of American society may be the number of citizens who find even this minimalist and relatively quiescent form of national government to be oppressive. The bombing of the federal building in Oklahoma City and the growth of armed "militias" are only the most extreme manifestations of a more general cultural phenomenon. Americans have historically been as much committed to their state and local governments as to the federal government, but the differential commitment and differential evaluation of efficacy of the governments has been tending to increase.

Norway. If the culture of the United States can be characterized as individualistic, then that of Norway can best be typified as egalitarian. That is, the society is characterized by high group values but low grid values. Individuals appear to identify strongly with groups in the society (albeit also with the country as a whole), and these groups help to organize a great deal of the day-to-day life in the country. That having been said, however, those groups do not present the average Norwegian with a predetermined set of rules to govern behavior, hence the low grid value. There does appear to be, however, a set of "meta-rules." Those meta-rules are that any group should be governed extremely democratically and consensually internally, and that groups should also interact with one another in an equally open and democratic manner in public settings. These meta-rules appear to extend to all sorts of organizations, including those that might be hierarchical in other societies. It is, therefore, difficult to identify any substantial patterns of hierarchical subordination in the society, also clearly a characteristic of Norwegian social and political history.[6]

The egalitarian principles that, we have argued, govern social interactions in Norway appear to have some anomalous features, as did the individualist characterization of the United States. The most significant of these is the importance of law in Norwegian government and society. The phrase "with law we will build the land" is as significant today as when it was first used. If an egalitarian culture is characterized by a low level of restraint by rules, then one might not expect law to play a prominent role in the governance of that society. In the case of Norway, there is a clear reliance on law as a framework of social action, as the training for civil servants, and as something of a totem for the dominant collective values.

There appear to be several ways to explain this apparent contradiction in the culture. One approach would be to argue that the legal framework in question is, as described above, largely a set of "meta-rules" that have their effect largely through setting the conditions for subsequent bargaining and consensus building. Law then serves as a set of external parameters or the "frame" (Schon and Rein 1994) that defines the space within which bargaining and compromise can occur legitimately. In the Norwegian context, that space for bargaining is rather extensive, and groups are granted substantial latitude to define the parameters of their own decision-making demands and acceptable outcomes. They are also granted substantial latitude in defining the nature of their interactions with other groups when involved in public decision-making roles (Kvavik 1976).

Following from the above, we can argue that law serves as a means of reducing transaction costs in this environment and therefore as a form of "compliance ideology" (Wilson 1992) in what might otherwise be a highly contentious setting.[7] This mode of analysis informs a good deal of institutional thinking in economics (North 1991) and can also assist us in understanding how policy is made through negotiations in relatively open political systems. If bargaining were to occur on a *tabula rasa*, then the need to redefine rules and the range of possible outcomes might present immense problems. Instead, law structures interactions within the institution and serves to simplify a bargaining process that might otherwise border on being impossible.

The other way out of this apparent contradiction over the egalitarianism of the Norwegian political culture is to remember that in high-grid culture, not only are there rules, but there may also be differentiated sets of rules for different social groups; for example, the caste system of India or apartheid in the former South Africa. This degree of differentiation would certainly not be encountered in Norway, and the same laws and rules would extend to all segments of the society in an almost radically egalitarian manner. Therefore public law becomes a means of ensuring that equality, and with that some of the Weberian elements of the administrative system that are crucial for guaranteeing equality of treatment.

POLITICAL CULTURE

The Douglas "group-grid" scheme provides a general approach to understanding culture. We will now proceed to discuss some cultural factors that are closer to affecting the political-administrative system and its decision-making behavior. Even with that greater proximity to the political-administrative institutions, however, there is some difficulty in establishing a clear link between those values and the behavior of political and administrative institutions, and of the individuals who occupy those institutions.

Attitudes Toward Legitimacy

An important factor in the political culture, especially for showing differences between the two countries under investigation here, is the concept of legitimacy and the source of legitimacy of government action. Legitimacy is the concept that government actions are accepted by the public as being right for them and that they therefore will be obeyed. Both of these political systems appear to be fully legitimate at the systemic level, despite significant if sporadic civil unrest in the United States. The principal comparative feature of this variable then is in the manner in which the two systems are able to attach the general legitimacy of the system of governing to the specific decisions taken within that system. We will be arguing that in Norway the principal means of legitimation is dependent upon the procedures of making rules, while in the United States substantive criteria of legitimacy tend to be taken into account, along with the important procedural elements.

Although they may be fully legitimate at the systemic level, the two systems may differ somewhat in the legitimacy attached to specific institutions of governing. In Norway, there tends to be generally strong support for all the institutions of government, with those institutions being relatively undifferentiated in the public mind. In the United States, however, there are marked differences in the legitimacy accorded different institutions. The public bureaucracy as a means of making authoritative decisions for the society has little support in the United States, unlike in Norway and most other European countries. Thus, although the American public bureaucracy is very active in issuing secondary legislation—regulations in the language of American government (Kerwin 1994)—there is a substantial literature questioning the legitimacy and desirability of this form of rule-making (Schoenbrod 1993).

On the other hand, the court system of the United States has a great deal of legitimacy—despite major controversies over decisions such as abortion and prayer in schools—whereas that is rarely thought of as an important rule-making institution in Norway. Indeed, the United States can be said to use the court system as a means of policy-making to an excessive and almost unwise degree. This pattern of making policy may invest relatively mundane matters of policy with constitutional significance that at once makes them more significant and more divisive than they might be, and makes subsequent adjustments to changing circumstances more difficult. Our recognition here of those negative features will not, however, alter the litigious nature of American society and American policy or the social system that throws up unbargainable issues of race, gender, and religion.

The government of Norway has a large reservoir of legitimacy among its citizens, but in order to be able to govern effectively it must still follow carefully the "rules of the game" in governing. These rules are not only the formal rules about how to make laws and regulations, but also include a variety of cultural understandings about how government should proceed when in the business of making decisions. In Norway, as in any parliamentary system, the government often has the formal capacity to push legislation through the Parliament, but the informal rules minimize any attempts at imposing this "majoritarian" style of governance (Lijphart 1991). The consensual style of governing makes such a show of political force unpalatable and perhaps unacceptable even to members of a majority.

This capacity to force legislation through is especially limited in Norway because of the frequent minority and coalition governments (Strom 1990). This form of government functions better than it might in other countries, but there are still impediments to its exercise of power. These governments are further restricted by cultural norms that, if a law is to become legitimate, it should be the product of extensive consultation in the society, with some notion that reaching a consensus among the affected parties is as important in the long run as making a decision quickly. Norway has a very large number of organized interest groups, despite its relative socioeconomic homogeneity, and these groups believe that they have the right to be consulted about issues as government makes policy. In some instances this right of consultation may be enshrined in law (Christensen and Egeberg 1979, 1997b; Kvavik 1976), but in others it may simply be an understanding among the participants in government about how the game of politics should be played.

In the United States, on the other hand, although there is certainly also a definite importance attached to procedures and due process, there is perhaps a stronger element of substance that is associated with the procedures. The Constitution does specify procedures, but it also contains a strong substantive element. In part, the politics of cultural pluralism mentioned above have tended to emphasize the need to examine the substance of decisions as well as the manners in which decisions were arrived at.[8] From the founding of the Republic, many decisions severely disadvantaging minorities were reached by completely legal procedural means, but the substance of those decisions has now come to be perceived as being inherently unfair and illegitimate and therefore not "due" in the current interpretation of "due process."

The emphasis on the substantive correctness of policy means, in part, that it is more difficult to impose differential costs of policies on groups in the United States than it might be in other countries, particularly Norway. The mere fact that a cost, be it a tax, the direct costs of a public benefit, or the siting of a nuclear waste facility, has been decided upon through appropriate procedures is of little consequence in a system that emphasizes individual liberties and substantive due process (Sunstein 1990). The more important element for most of the parties concerned is the substantive content of the decision itself. The American political system provides a number of alternatives for procedural process and legitimation and a number of different governments through which to legitimate a decision, so that any one decision is just that—one of many possible legitimate decisions—and may be just the beginning rather than the end of the political process.

The structural elements of American democracy tend to make this concern with the substance of policy all the easier to put into action. In the first place, it is difficult for government to make decisions that might disadvantage anyone, given the level of political organization and the direct access of most interest groups to government actors (Laumann and Knoke 1987). The conventional conception of "iron triangles" having policy dominated by a single group (usually a producer group) with all others excluded is now rather antiquated, although it may still exist in a few policy areas. The more common metaphors now are "communities" and "networks" (Laumann and Knoke 1987), implying a wider exercise of access by interest groups, and a less-determinate policy outcome than would be true for the iron-triangle characterizations of American politics.

Although the network metaphor may be applicable to a wide range of policies, not all networks function in the same manner. For example, Sabatier (1989) and Jenkins-Smith and Sabatier (1993) have focused on those instances in which opinions within networks are divided. In these cases, the participants tend to engage in conflicts based largely on the marshaling of scientific evidence and attempts to encourage the other side to "learn" about the virtues of one or another policy position. Other network descriptions tend to have a more unified view of how they function, even if there is a common focus on learning

(Adler 1992; Haas 1990). Also, subnational governments may enjoy a stronger position in some networks (e.g., education) than they do in others, and the nature of intergovernmental politics, rather than of interest groups themselves, may dominate patterns of behavior with those networks.

The network metaphor is applicable to the access of groups to administrative agencies as well as to Congress. For example, if the administrative agencies want to make a new regulation, they are required to open themselves up for consultation and to pay some attention to the thoughts of the segments of society affected by the proposed regulations (Bryner 1986; Kerwin 1994). Interests do not necessarily win through their involvement with the regulatory process, but they do have the right to be heard. Further, interest groups have been able to gain access to the legal process as well, largely through the use of class-action suits and amici briefs (Caldera 1992), and thus can attempt to prevent their members from suffering deprivations from court decisions. Again, using any of these points of access will not guarantee that interest groups will win, but the multiple decision points provide them more opportunities to produce, and especially to block, actions by government agencies.

Attitudes toward Law, Judicial Expertise, and Due Processes

The cultural traditions of legitimacy in political-administrative systems and the emphasis on procedure and substantive due processes in Norway and the United States, respectively, lead us to discuss a related cultural variable—namely, attitudes towards the judicial premises in the political-administrative process.

The courts in the United States serve as the principal arbiters of what is or is not substantively correct in decisions. Beginning as early as at least the 1930s, the Supreme Court began to develop the concept of substantive due process to supplement the more obvious criterion of procedural due process derived from the Fifth Amendment of the Constitution. In a series of cases, the Court argued that the constitutional admonition that government should not deprive citizens of life, liberty, or property without due process of law means that the substantive criteria by which the deprivation is made must be legitimate, just as are the procedures. This restriction is just as much as must the process be by which the decision about the deprivation was made. Any differentiation of citizens in law made on the basis of ascriptive criteria such as race or gender, for example, tends to be suspect immediately, whereas other differentiations made upon the basis of behaviors tend in general to be more acceptable.[9]

Related to the concept of due process, the political culture of the United States is highly legalistic. There are many more lawyers per capita in the United States than in Norway, or indeed in almost any other country in the world. It is now popular among Americans to condemn lawyers and to point to their excesses as a root cause of many public problems, including the rising cost of medical care and the relatively poor competitive position of American industry. On the other hand, the same people who condemn lawyers in general are more than willing to consult a lawyer themselves if they believe that some other actor (including government) has infringed on their rights. They are particularly anxious to consult a lawyer if think they can receive a good financial settlement. Thus, for Americans, the legal process has become a generalized means of conflict resolution for the society, while for Norwegians the law is a much more particularized mechanism and not at all that important in political-administrative processes. Both societies have a well-developed conception of the rights of citizens, but in Norway political and administrative action, rather than legal action, is generally perceived to be the more appropriate mechanism for confirming those rights.[10]

The questions of legality and of authority point to the question of the diffuse or specific support that citizens may express for their governments in these two countries. Again, both systems have strong diffuse support, but systemic support appears even stronger in Norway than in the United States. One way to assess this factor is to look at the confidence and trust that citizens express in their government and its institutions. The level of systemic support is substantially higher in Norway than it is in the United States. Further, that support tends to be spread across the entire range of political institutions, while positive affect in the United States tends to be expressed toward a relatively few public institutions. In particular, the American legal system and the police tend to be rated rather highly, while Congress and the bureaucracy are rated extremely poorly (*Gallup Monthly Report* June 1996). The American reaction to government appears to be more specific and performance related, despite the high levels of patriotism found in the society.

Not only are the confidence and trust of citizens in their government important indicators of diffuse support, but also their behavior vis-à-vis the rules and laws of their government indicates that support. Not only should citizens obey the laws if they have diffuse support for the system, but they should also do so out of general respect for those laws and the processes that produced them (Rogowski 1972). The evidence from social surveys is that obedience to the law is a complex question: Some of the decision to obey is a function of the fear of being caught and punished (Spicer and Lundstet 1976). Another part of the obedience decision reflects acceptance of the mechanisms through which the laws are made. Finally, another part of the willingness to obey results from diffuse support for the political system and for the rights of government to make decisions in the name of the national community.

It is difficult to address the question of the sources or the comparative level of obedience to laws directly. Most respondents to surveys are reluctant to say that they have not obeyed laws or to differentiate among the various reasons why they would obey or disobey the laws. Still, there is some evidence on these points. Some surveys concerning tax evasion have found that, for tax laws at least, there tends to be relatively little feeling that compliance is obligatory by virtue of being a citizen (Laurin 1986; Listhaug and Miller 1985). Rather, citizens tend to be very pragmatic about compliance. They do recognize that government does have some claims on their resources, but they see no reason to comply to the fullest. To the extent that they do comply fully they do so as much out of fear of the consequences as from a deep commitment to paying taxes.

There are some cross-national differences in citizens' reactions to questions of compliance with tax laws. As might be expected, citizens of southern European countries tend to have lower levels of compliance and less feeling of a moral obligation to obey tax laws (Peters 1991). In this regard, Americans and Norwegians do not tend to be very far apart, despite the findings on other indicators that the Norwegian government has a larger pool of diffuse support than does the American government. In this instance it may be that the substantially higher tax burden on the average Norwegian is translated into a belief that a certain amount of discrete tax evasion may be worth the risk and is not really a crime.[11]

The limited evidence concerning compliance with other types of law is somewhat more comforting for believers in diffuse support for political systems. Citizens have a direct and measurable economic interest in evading taxes. Also, that behavior may be a means of protesting against perceived government excesses without really undermining the legitimacy of that government. Ignoring or wantonly disobeying other laws may be more threatening, both to government and to the average citizen. Tax evasion may be acceptable to many people in high-tax countries, but disobeying other laws is perceived to be a more serious affair. This may be true even for seemingly minor offenses such as speeding and not wearing seat belts.

Political Trust

We mentioned above the concept of political trust while discussing the concept of legitimacy and authority of governments. It is certainly a crucial element in that fundamental relationship between state and society, but it also has a number of other important influences on the conduct of politics in these and other societies. This variable of trust is far from undifferentiated and should be broken down into two basic questions: First, do the people have basic political trust—do they believe that the occupants of political roles are honest and working for the common good? Second, who do the people trust? If we begin to answer the second question, then the answer to the first will become apparent.

Objects of Trust. When the population surveys its political and social environment it is faced with a number of individual and institutional actors. Many of these actors average people may ignore in their day-to-day lives, but they may still have some opinion about their trustworthiness. These images may be the products of the media or in some vague sense are derived from the experiences of others, but they are nonetheless real perceptions and may affect behavior. The average citizen will tend to seek out individuals and institutions in which he or she has some trust, and to avoid those that are perceived to be less trustworthy.

The first objects for trust that we will discuss are the fellow citizens of any other citizen. In other words, does that average American or Norwegian believe that his or her compatriots are worthy of their trust, or do they believe that those other people are only out for their own personal good and are not trustworthy? We can discuss interpersonal trust generally (Fukuyama 1995), but we can also look at it more specifically in regard to politics. Do my fellow citizens recognize my political liberties and respect my right to express my political opinions? Are those other citizens involved in political action just for their own good or are they motivated (at least in part) by a sincere interest in the common good? And do they have something of the same conception of that common good that I do?

Clearly, a society in which the majority of people did not believe that their fellow citizens were engaged in politics for reasons of mutual benefit for the society would find political organization and collective political activity difficult. Fortunately, neither Norway nor the United States falls into that skeptical category. In both countries there is a general sense that their other citizens are relatively benign and worthy of trust. This interpersonal trust manifests itself in a number of ways in addition to being just the response of people to survey questions. One of the most important of these is membership in politically relevant groups of all types. Robert Putnam (1993) found this variable to be very important in explaining differences in success in governing in different parts of Italy; areas with a large number of private organizations seemed more able to govern themselves than could those with a less vigorous organizational life. Putnam has argued that social capital is also becoming more problematic for the United States, as more people are "bowling alone." The same effect of trust on performance should be expected in cross-national research.[12]

If a citizen does not believe that other people are to be trusted, then he or she is very unlikely to join with them in political parties or interest groups. The United States and Norway are among the more organizationally intensive countries in the world, and in that they demonstrate a marked contrast with some other industrialized democracies such as France and Italy. In both countries there is at least one organization (and usually several more) for almost every cause and purpose, and people appear to encounter few impediments to joining with their fellow citizens for political purposes. This level of interpersonal trust is especially important for the United States, since it provides a secondary means of governance for a society that is difficult to manage through the formal institutions of government.

Political and Administrative Institutions. As well as interpersonal trust, trust between citizens and political institutions is also important for providing effective governance. If the population believes that the individual and institutions responsible for governing are worthy of the trust of citizens, then they will be willing to provide those institutions with a certain amount of latitude for action. That latitude for action, in turn, may produce greater efficiency and effectiveness. There is a clear contrast in the level of trust in political institutions in the United States and Norway.

In general, Norwegians do possess a high degree of trust in their political institutions, as well as in the political leaders who work within those institutions. Respondents in surveys in Norway report extremely high levels of confidence in their political leaders, especially when their responses are compared with responses in most other countries—even other industrialized democracies. This political confidence expressed by Norwegians extends across the range of political institutions, including the public bureaucracy in which relatively few other populations express much confidence or trust. This high and pervasive level of political trust provides government institutions in Norway with a major reservoir of legitimacy that can be used to make more difficult policy decisions than might be possible in the United States (other than through the court system perhaps).

Respondents to surveys in the United States, on the other hand, demonstrate a low and declining sense of trust in their political institutions and their political leaders. There is, however, some differentiation in popular beliefs about government institutions in the United States. The courts and the police are still rated relatively positively, while Congress and the executive branch of government are rated extremely poorly. The popular evaluation of the presidency has tended to be highly variable, depending upon circumstances and the performance of the president in office at the time.

The popular evaluation of the presidency appears to some extent to be a surrogate for their perceptions of the performance of the entire political system, as well as the economic performance of the system. The bureaucracy tends in general to be negatively evaluated at virtually all times. Despite that generally negative assessment of the bureaucracy, the assessment of individual programs and contacts of the public with the bureaucracy tend to be more positive (Goodsell 1995). This schizophrenic response of the public to government organizations and employees is not dissimilar to their assessment of Congress—despising Congress but having a great deal of affection for their own congressmen.

To some extent the differential between the United States and Norway in popular trust and confidence may reflect some actual differences in the manner in which government has been conducted in the two countries. At the level of public policy, the extensive welfare state in Norway provides a range of social and economic services not available through the public sector in the United States. These services are bought at a much higher tax price in Norway than in the United States. Still, the public sector can be considered as a benevolent and, for the most part, efficient provider of services to citizens, while that of the United States remains a laggard.

The assumption that these public services are conducive to political trust, of course, depends upon their being positively valued by the public, something that may be more true in Norwegian politics than in those found in the United States. Indeed, government in the United States is appearing to gain greater trust by reducing the little welfare-state programming that does exist. There is, however, an interesting question of learning and acculturation here: Had Americans become more accustomed to an activist and more benevolent government earlier in their national development, would they too equate good government with a more activist version of governance? Or would the cultural imprint of the founding of the Republic have persisted, with its rejection of any centralized power and its cult of the rugged individualist? This is, of course, impossible to tell, but it is clear that this did not happen and the United States persists in its antigovernmental political culture.

Such a benevolent characterization cannot be argued to be true of government in the United States, especially the federal government. The welfare state provisions in the United States have expanded over the past several decades, but still lag well behind other industrial democracies (Castles 1993). State and local governments have a more substantial social-policy profile, but the federal government still is primarily engaged in the business of defense and a few social-insurance programs. The 1994 Congressional debate concerning the creation of a national health-insurance program is indicative of the continuing absence of a strong social-policy commitment in the U.S. federal government. Additionally, the overwhelming victory of the Republican Party in the 1994 Congressional elections has tended to reduce further the federal role in social policy and indicates the importance of "right-wing populism" in contemporary American political culture (Kazin 1995).

Similarly, the style of governing in Norway is more likely to generate trust in the political system than is that of the United States. Although surely strained at times, consensual norms are still extremely important for determining the manner in which policies will be made and then administered in this system. This reliance on consensus means that policy-making will often be slow and deliberate, but also that its outcomes will almost certainly take into account the full range of interests and ideas in the society. Americans often believe that their views have been excluded from decisions in Washington (or even their state capitols and city halls), and often vote to express their discontent with that perceived exclusion (the November 1994 elections, for example). Norwegians, on the other hand, have a more difficult time believing that their views have been excluded from decisions made in Oslo.[13]

The analysis of exclusion and decision-making costs in government put forward by Buchanan and Tullock (1962) appears directly applicable in this comparison. The recent pattern of governing in the United States is no more conducive to trust than are the substantive outcomes. Governance has been described as "gridlock" (Levine and Kleman 1993), "divided government" (Fiorina 1991), and by even less complimentary terms. While some scholars argue that the institutional and political divisions within American government do not prevent effective governance (Mayhew 1991), the average American finds the "squabbling" that is seen each day on the evening news as evidence of the incapacity of their current governors to do that job effectively. While Americans generally laud the institutional framework for policy-making created by the Constitution, and the individual congressmen within the structures, they are simultaneously extremely discontented with the actual workings of that system.

Interestingly, one of the most common reactions of critics of the contemporary system of governance in the United States is to make it less manifestly democratic. The recommendations for making American democracy function better often involve substituting formulae, constitutional rules, and automatic processes for open democratic processes of decision-making (Hanuschek 1986; Weaver 1989). Given that the strictly democratic processes of legislating often must confront an absence of sufficient political will to control public spending and the creation of entitlements, automatic controls are proposed. Mechanisms such as the Gramm-Rudman-Hollings procedure, the Budget Enforcement Act of 1990 (Kettl 1992), and the (often proposed but difficult to enact) balanced-budget amendment (Aaron 1994) are all designed to protect politicians from themselves and to force them to be more responsible.[14]

In addition to a higher level of general trust in government, Norwegian government benefits from the legalism of the system. The integrating force of law permits lower levels in the bureaucracy to be empowered and for there still to be a clear rule of law. The greater variability of educational backgrounds and the multiple criteria (law, economics, management) in American bureaucracy (Moe 1994) will make it more difficult to ensure a common interpretation of and obedience to law in the American setting. This is true despite the widespread respect for the legal framework created by the Constitution. In the instance of most civil servants, legalism and constitutionalism in the United States are often more a part of the common myths of society, rather than immediate operational restraints on the behavior of individual decision-makers.

SPECIFIC CULTURAL VARIABLES

Culture Related to Institutions, Policy Fields, and Professions

Selznick (1957) emphasizes that cultural norms and values develop gradually according to internal and external pressure, in a process of institutionalization. Political-administrative systems, interinstitutional structures, institutions internally, and certain political and administrative roles are all getting certain features that makes them unique, regardless of similar structural features. These cultural norms and values will in certain systems, institutions, roles, and time periods be compatible with the formal norms, but can also counteract these and have a stabilizing or obstructive function. Krasner (1988) stresses that "the roads taken," the resources invested, the contacts made, and the knowledge involved make it difficult for institutions to make choices and act in ways that is deviant—that is, they are "path-dependent." This implies that understanding what institutions do involves understanding the culture-specific reactions.

Our analysis of the structural patterns of the systems in Norway and the United States, and the related cultural features, illustrates this main point. The macroinstitutional design in the United States, stated in the Constitution, creates a fragmented structure both between and inside main political and administrative institutions. This structure and its development are partly conditioned by and partly participate in developing a culture that is characterized by heterogeneity, conflict, and competition. In the Norwegian system, the parliamentary principle develops a culture characterized by homogeneity, cooperation, and low tension. The structural and cultural norms in the two systems seem evidently to be reinforced. But such reinforcement processes are also modified in the two systems, even though this is not easy according to the cultural argument. Central political and administrative actors in the United States also develop cultural patterns that further cooperation and deliberative processes, and central actors in the Norwegian system are struggling with fragmentation and problems of capacity and coordination.

The argument stated above is also a general one applying to policy areas and professions. One important source of differentiation appears to be the policies being decided upon and administered (Freeman 1985). Health issues, for example, are influenced by the cultural values and practices of the medical profession, even when policy decisions are being made by individuals who are not themselves physicians (Friedson 1986; Immergut 1992). Further, individual organizations develop their own internal cultures (Ott 1989; Peters and Waterman 1982) that influence the behavior of the members of the organizations, as well as the overall pattern of decisions made. Even in organizations such as ministries of defense that presumably are well integrated, there is evidence that the organizational cultures of their components influence decisions and generate conflict and competition (Allard 1990; Smith, Marsh, and Richards 1993).

At times, different organizations within the same country responsible for much the same policy area will develop very different ideas about policy and administration. For example, antitrust issues are handled by both the Federal Trade Commission and the Department of Justice in the United States (Eisner 1991; Peters 1999). The former body has tended to adopt an economic perspective on these issues, while the latter has a more strictly legalistic perspective, and the two organizations tend to focus their efforts on different types of cases. Both organizations employ both economists and lawyers on their staffs, but differ in how their internal organization and their histories emphasize one set of values or the other.

Similarly, in Norway, responsibility of dealing with refugees is divided among a number of organizations, each imposing its own definition of the issue. The development of the oil resources in the North Sea has been characterized by competing definitions, one emphasizing technical and economic arguments, another taking a more conservationist view. A similar conflict has been evident in the development of the hydroelectric power in Norway. Yet another example is the discussion about how to organize foreign aid in the governmental structure. An independent solution is favored by the experts in the field, often more "client-oriented" than actors that favor more political or economic premises in this policy area, and therefore wish to integrate the policy field either in the Ministry of Foreign Affairs or close to the foreign-trade administration.

Organizations also are embedded within other, larger organizations, and those superior structures too can have an influence over their values and performance. For example, in the United States, the Forest Service is a part of the Department of Agriculture, rather than the Department of the Interior in which the National Park Service is housed. Not surprisingly, the National Forest Service has a greater orientation toward the production of forest products than would be expected if it were in Interior, where it might be expected to have more of a conservationist orientation. Likewise, when an agency is moved from one Department to another it may have to respond to a different set of signals from a different internal culture. The shifting of an organization from one setting to another is more than a structural manipulation; it also can be shifting of cultures and values. The frequent movements of the U.S. Coast Guard are indicative of the potential impact of organizational cultures; it has been part of the Treasury Department and the Department of Commerce, as well as its current location in the Department of Transportation. It also has been part of the Department of Defense in time of war.

For Norway, the most interesting case of a superior level of government or organization influencing behavior is its relationship with the European Union (EU). Although Norway has chosen twice not to join the EU, it cannot escape the influence that Brussels and its programs are having on political life and public policy throughout Europe. There is evidence that this relationship with Europe has forced some changes within the Norwegian administration. Over a long period of time, Norway has adjusted to the central norms and values of EU, especially sectorial ones, acting nearly "as if" it were a member. Studies of the adjustment of the civil service to the EU in the late process of applying for membership (and say "no" in a referendum in 1994) show clearly that the adaptation is very much based on the existing cultural–institutional norms and values (Christensen 1996; Farsund and Sverdrup 1994).

Cultural Elements of Public Management

We will then discuss the specific cultural elements that surround public management. Managing a public organization and private organization presents some similar challenges and will be influenced by some of the same attitudes about factors, such as hierarchy and interpersonal relationships, that exist within any organization. Of course, there are a number of features that differentiate public- and private-sector management (Allison 1987; Rainey, Backoff, and Levine 1976). These factors may, however, have as much to do with the content of the decisions taken as the manner in which people actually work together within an organization. In particular, the issues that arise in the public sector are not always subject to resolution by the economic logic of private-sector management—despite what neoliberals in the public sector currently espouse.

This culture of public management is in part derived from general ideas about governing, and in part derived from general ideas about organizational management. Public administration is clearly a part of government, but it is also influenced by ideas coming from business. Indeed, the spread of the private-sector ideal in most countries has made public administration even more subject to the thinking in the private sector, as has the increased movement of personnel between the sectors.

The so-called New Public Management movement in many countries can be defined in different ways: One interpretation is to emphasis that government and public administration must be less expensive and more effective, and therefore apply principles from the private sector. Another is that this "concept" is first of all supporting norms and values that emphasis economy too much, and that this is threatening other legitimate and democratic values and norms in the political-administrative system (Christensen and Laegreid 1996). The concept is also seen as too ambiguous and unable to give more specific direction to governmental action, and is therefore often characterized by symbolic manipulation (Christensen 1991b). Is the former point of view characteristic for the culture in government in the United States, and the former one typical for Norwegian reactions to this new reform wave?

One strand of New Public Management elements in the United States in thinking and action is the current spate of "reinvention" and "empowerment" exercises. Associated with these are analogous ideas of "customer service" in the public sector. These changes are commonly associated with the Gore Report (1993), but are manifest in a number of other attempts to create change within the public sector. The basic ideas are to make government less bureaucratic (in the pejorative sense) and to enable lower-level officials to make decisions. As has been noted in a number of places (Kernaghan 1992; Pierre 1995), while these reforms may solve one of the perceived problems of government, they may exacerbate others. American government may be able to make decisions more quickly, but those decisions may be less accountable and less governed by the rule of law. Ultimately, then, these reforms may make government less legitimate in the eyes of many citizens.

The apparent trade-off between efficiency and the rule of law does not appear to be as dominant in Norway as in the United States. Norwegian government appears capable of acting decisively through the public bureaucracy, while retaining the impression (and usually the reality) that there is a rule of law. This appears possible for several reasons, mostly occurring higher up in the hierarchy of values we have been discussing in this chapter. In particular, the general level of political trust existing in the Norwegian system enables it to function with fewer manifest external constraints. The apparent need to "deregulate government" in the United States (DiIulio 1994) is one indication of the multiple levels of control that have been built up in the system, in large part in reaction to popular distrust of government in general, and especially of the federal government (but see Barzelay 1993).

We could not argue that Norway has been entirely immune to the "fad and fancy" surrounding reform of the public sector. Many of the same issues of reform that have at least been broached in Oslo have been implemented in London, Canberra, and Washington, even though implementation comparatively appears culture-specific (Olsen 1995b). Still, with all that discussion, the reforms themselves and the manner in which they have been introduced are by no means as radical as in Anglo-American democracies (Christensen and Laegreid 1996). There is a sense that the system of governing "ain't broke" so there is no reason to fix it. Further, the reforms that have been implemented, especially in the Anglo-American systems, are too much in the market-and-individualistic model for a system with a more communal political culture.

Cultural Elements in the Individual Roles

Cultural factors are politically significant through the roles that individuals choose to play in their public behavior, whether they are political or administrative. Culture also comes into play through individual perceptions of the socioeconomic environment and of policy problems. For the individual decision-maker in government, his or her role behavior is influenced by all levels of culture mentioned above, although *ceteris paribus*, the lower levels of culture will probably have a more proximate impact on behavior than will the more remote level of national culture. For example, in their day-to-day lives, the occupational or organizational roles of political actors will have a greater impact on decisions than will the general roles derived from national political values.

First, we will emphasize the degree of consistency in individual public roles. The relevance of consistency to decision-making and governance can be that more integrated and unambiguous role-anticipation makes it more easy to act in accordance with public goals. Is this consistency lower in the United States than in Norway? One main argument supporting the view that the political or administrative roles of actors in the United States are characterized by a culture of inconsistency is that the political-administrative system is more fragmented, one has to attend to many different considerations at the same time, and there is a complex set of rules to attend to. In Norway, there are fewer and more agreed-upon considerations to attend to and therefore is easier to socialize new actors into their roles and create consistency (Christensen 1991a; Eriksen 1988b; Laegreid and Olsen 1978).

Second, the exposure of and definition of accountability for politicians and bureaucrats may vary between the two political-administrative systems. Exposure and accountability can on the one hand be seen as a democratic safeguard, as part of political and popular control, but on the other hand, make political- administrative systems vulnerable to resourceful actors and symbolic actions. Generally, politicians and bureaucrats in the United States seem to be much more exposed to public accountability and held more personally responsible, on a judicial basis, for decisions and outcomes than in Norway. Central political and administrative actors in Norway are characterized by being far less exposed and more collectively responsible, derived from elected, public bodies.

Third, the discretionary influence of public-sector bureaucrats on decision-making may be enacted in different ways in the two countries. Discretion is, on the one hand, one way to handle problems of capacity and increase flexibility and decision quality in a political-administrative system, but may also potentially undermine the political control and increase the influence of administrative actors. The discretionary autonomy seems to be more narrow and the political premises more important in the United States than in Norway, where formal autonomy is higher and the professional considerations play a more important part in discretionary behavior. In Norway, the discretion of the civil servants has increased, especially after World War II, without threatening democratic control, since there is evidently some shared cultural norms and values between politicians and bureaucrats (Christensen 1991a; Christensen and Egeberg 1997a). In the

United States, there seems generally to be a tendency in the same direction, but it is modified by a strong increase in the means of control towards the civil service, as discussed in chapter 5 (Kerwin 1994; McGarrity 1991; Schoenbrod 1993).

Conclusion

Our main conclusion is that the United States and Norway first of all have cultural features that are quite different and seem to sum up into different cultural types, even though there also are similarities and modifications of these types. The United States is generally characterized by cultural heterogeneity (while Norway is very cultural homogeneous), a feature that seems to create extra load on the decision-makers. In trying to cope with this heterogeneity and load, the central decision-makers in the United States must cope with many obstacles: One is the individualism in the culture; another is the substantive due processes; a third is the low trust in political institutions—all features that may make political-administrative coordination and control more difficult. Adding to this are the conflicts between and inside the executive and legislature and the low trust in the bureaucracy, leading to ineffective decision-making processes and tight and complex control of administrative units and roles.

But these cultural features of the system in the United States can also be evaluated differently. A fragmented, conflictual, and individualistic culture can reflect a dynamic system that is attending to different complex interests and considerations in open and democratic processes, and thereby securing its legitimacy. A legalistic orientation can both secure individual rights and safeguard the democratic control of political and administrative actors.

The homogeneity, lack of conflict, egalitarian values, and trust in political institutions in Norwegian culture makes it probably more easy to reach collective decisions and implement them than in the United States, but also creates a centralized system that can have problems with attending to differentiated interests and needs. The procedural orientation, the meta-rules, and lack of legalistic orientation may make it easier for the central political actors to control the decision-making processes and secure legitimacy, but also lead to problems in securing individual rights. Trust in political institutions and giving central actors high discretionary influence may secure flexibility and effectiveness in decision-making processes, but also make institutions and central actors insulated from critical and public knowledge about the content of decision-making processes.

If we return to the central theme of this book, we can see how culture has interacted with structural features of both government and society, and tended to reinforce the impact of those features. The more fragmented and distrustful (of government and increasingly other citizens) culture of the United States helps to reinforce the fragmentation and blockages endemic in the formal political structures. The inability of the system to process demands effectively and efficiently, in turn, reinforces the belief that government is incapable of doing anything well and that the private sector—usually business but also the voluntary sector—is the only mechanism capable of actually providing direction to the country. Of course, what that vision of governing implies is that there will tend to be many directions taken, rather than a single coherent stream of governance. The system remains, as Richard Rose (1980) argued, little more than a series of subgovernments.

Norway, on the other hand, demonstrates a more "virtuous circle" in the reinforcement of its cultural and structural features. The political culture and mass society certainly do contain substantial politically relevant cleavages, but they also display a substantially higher level of integration than those encountered in the United States. Further, the political-administrative system of Norway is relatively well integrated and capable of acting promptly and effectively and, for the most part, speaking with a single voice. This integration is found even though there is apparently a good deal of fragmentation in the structures of the bureaucracy, with the agencies having some degree of latitude for action. This fragmentation may be more apparent that real, however, and the system is well coordinated through both legal and hierarchical means. In short, the Norwegian political system is capable of delivering a more coherent and integrated form of governance.

We may, however, have been applying an inappropriate standard to governance in the United States. In many ways, the American system has been extremely effective in delivering the type of government favored by the framers of the Constitution, and is still favored by many Americans. That is, the system has been designed extremely well to prevent rapid, coordinated action except in times of extreme duress—wars or other major emergencies. Further, it permits the states and localities substantial latitude in pursuing policy goals and policy means that the central government could not impose upon the entire society. While it appears terribly incoherent to the rationalist observer, in its own terms the system is extremely successful.

NOTES

1. This phrase is not constitutional but rather comes from some of Thomas Jefferson's correspondence. It has, however, been picked up by first amendment scholars and some jurists to describe the meaning of the constitutional phrase concerning establishment of religion.

2. This movement is largely fundamentalist Protestant, but also draws some strength from conservative Roman Catholics, especially on the issue of abortion. Some Orthodox Jews also have made common cause with this movement over the increasing secularism of American society.

3. Perhaps the major exception might be the continuing pressure to equalize the role of women in society and the relatively minor problem of the indigenous Sami people.

4. Despite the fears of advocates of an "English-only" policy, immigrants (or at least their children) quickly adopt English.

5. The logic is that if the group must make decisions by consensus, then bargaining and compromise with an outside group is likely to upset a hard-won unanimous position. As Buchanan and Tullock argued (1962), the decision-making costs of consensus are extremely high.

6. Unlike most of the rest of Western Europe, Norway did not have a feudal social structure during the Middle Ages but rather maintained some democratic institutions, even during that age of extreme hierarchy.

7. The base of contention would be the numerous divisions in society rather than conflictual behavior by the actors involved.

8. Constitutional law has come to separate "substantive due process" from "procedural due process." A series of Supreme Court cases established that no matter how a law or regulation was made, it could not be "due" if its substance violated conceptions of fairness.

9. A major exception would be programs that attempt to rectify severe racial imbalances in activities such as higher education. The debate over scholarships targeted for minorities is a recent example.

10. There is a legal mechanism somewhat like judicial review in the Norwegian system, but it does not have nearly the power to forestall government action as would be true of that instrument in the United States.

11. In neighboring Sweden, respondents to a survey once argued that tax evasion was not a crime, but merely a necessity. See Vogel (1974).

12. For a rebuttal, see the *Public Perspective* (June 1996).

13. The analysis of exclusion and decision-making costs in government put forward by Buchanan and Tullock (1962) appears directly applicable in this comparison.

14. State governments in the United States, reflecting public sentiments perhaps somewhat better than the federal government, have balanced-budget requirements in state constitutions (with one exception). Many state constitutions also require a public referendum before the government is allowed to issue any bonded indebtedness. Also, several major cities have had to place their budgets under "boards of control" that lack any political base.

CHAPTER SEVEN

Governance in the Two Societies

Main Perspectives and Empirical Results

This book has focused on structural and cultural features of the political-administrative institutions on the central level in the United States and Norway, and their significance for the decision-making processes and governance in the two systems. Our theoretical point of departure was outlined in two perspectives, a structural–instrumental and a cultural–institutional one, the former emphasizing the importance of formal structure and organizational design and the latter stressing the significance of cultural norms and values. We anticipated that the analysis would show that structure and cultural features would be reinforced in the two system—in a heterogeneous and fragmented direction in United States and in a homogeneous and integrated way in Norway—and our main results seem to have confirmed this.

The emphasis of the empirical analysis is the structural features of the executive and legislative powers on the central level. Chapter 2 discussed the macro-political structure of the two systems, related to the presidential and parliamentary systems, and underscores the fragmented and competitive nature of the U.S. system and the cooperative and integrated nature of the Norwegian. Chapter 3 analyzed the development and contemporary internal structure and demography of the legislatures in the two countries. The two legislatures have faced some of the same problems of overload and capacity, and reacted similarly with increased specialization and growing staff resources. But Congress is generally much more specialized, fragmented, open for individual political efforts, less hierarchical, less party-based, and more heavily staffed than the Norwegian Storting, where the party is a centralizing and coordinative force. We also showed that Congress has many more instruments of control and points of access to the executive than in Norway, where there are few shared functions and the Storting does not interfere much in the daily operations of the executive.

Chapter 4 analyzed the structural development of the executive in the two countries, stressing that, over time, the executive in the United States has built up a large, specialized apparatus close to the president, and that the cabinet as such never has been any important and collective political force. In Norway, there has always been much reluctance concerning building up resources close to the prime minister, but there is a long tradition in the political significance of the cabinet as a collective body. In the United States, the president has many different means of control over Congress, while in Norway the executive has influence over the Storting through the preparation and implementation of policies and programs, but on the other hand is closely responsible to the Storting and shares its central goals and values. In chapter 5 we analyzed the internal structure of the executive bureaucracy, emphasizing how much more fragmented the system in the United States is, consisting of many different organizational forms and liable to be influenced and controlled by individual politicians, Congress, the president, interest groups, actors on lower levels, and so on—more so than the Norwegian central bureaucracy, which is primarily much more influenced by the cabinet and its political leadership.

The second set of variables discussed has been the cultural ones. In chapters 2 through 5 we briefly discussed the types of cultural features that seem to have developed in relation to the main structures in the two political-administrative systems, and concluded that the cultural tradition in the United States is one characterized much more by individualism, competition, and conflict than in Norway, where collectivism and cooperation has been favored. Then we discussed in chapter 6 the different ways to support this cultural analysis. In so doing we selected three types of nested sets of cultural variables: First, factors showing cultural features on the systemic level. We showed that the United States is having many deep and cross-cutting cleavages that affect the decision capacity and effectiveness of the system, while Norway is much more socially homogeneous. The group-grid theory applied to the two system underscored the individualist orientation in the United States system and the egalitarian approach in the Norwegian. Second, we discussed some main features of the political culture, stressing that the United States is very much characterized by a legalistic orientation, substantive due processes, and low trust in the executive and legislature, while Norway typically has a procedural focus, meta-rules, and high trust in central political institutions. Third, we showed that the cultural norms surrounding individual roles of politicians and bureaucrats in the United

States are more inconsistent, exposed, and programmed than in Norway, where high discretion, low exposure, and high role consistency are typical.

On Governance

We will now return to our fundamental dependent variable in this analysis of politics in the two countries: *governance.* This term is employed with increasing frequently in the social sciences (Kooiman 1993), but it still has several alternative definitions, or at least there are several different components within the one broader concept. While the openness of the definition is to some extent useful and a way to forestall premature conceptual closure, it can also create substantial confusion. For example, in some more traditional views, governance implies a top-down, centralized control of society by the public sector, while in others it implies a much looser steering role, manipulating incentives rather than issuing commands (Kooiman 1993; March and Olsen 1995; Rhodes 1997). Both of these views have substantial validity, but they would argue very different things about the governing capacity of these two, or any other, countries.

As we will point out, that confusion is exacerbated by very different conceptions of the public sector and therefore about governance in the two societies with which we are concerned. For Norwegians, despite their strong commitment to democratic norms, there is a historically rooted willingness to accept a strong directive and redistributive role for the public sector. In the United States, despite pride in the political system and an equal commitment to democratic norms, there is much less willingness to permit government to have a significant role in steering society. Thus when we discuss governance, we must always be cognizant of the barriers to governance that may shape practice within the two systems, and especially in the United States.

A definition of governance is also made more problematic by the conflation of both normative and descriptive elements within the one term. The use of the word "governance" tends to imply that some persons or organizations have been successful in steering society in a certain direction[1]; it is usually a "top-down," normative vision of how government should function (Rose 1976). Thus governance tends to become defined as a positive social value, and consequently those elements of society that may obstruct governance tend to have a negative value attached to them—for example, the reputed *incivisme* of many southern Europeans (Seigfried 1940). This negative inference may be made even if those individuals or groups are operating through perfectly legal avenues and are merely exercising their legal and political rights—for example, in legal tax avoidance (as opposed to illegal evasion).

Thus, a focus on governance runs the risk of begging the question of whether the directions in which society are being pushed are desirable or not, and if so, for whom? Further, for more individualistic societies such as in the

United States, such a "state-centric" (Evans, Rueschmeyer, and Skocpol 1985) conception of the good society might never be acceptable. Similarly, the more egalitarian and collective Norwegian society might find the market-based ideas of much contemporary government-reform antithetical, even if government could be said to be governing when it imposes those reforms. Thus, although it is a social good of sorts, governance is not entirely neutral and we must therefore ask, *governance about what?*, and with what content to be able to assess performance adequately?

There are both structural and cultural elements involved in governance. On the one hand, political systems that get the machinery "right" (meaning appropriate for the tasks at hand), are more likely to produce successful governance than are those who have been less skillful or less fortunate in designing their machinery of government. The contemporary debate over presidential and parliamentary government, for example, assumes that institutions *do* matter (Weaver and Rockman 1993). On the other hand, governance is also cultural, with different conceptions of the appropriate nature and amount of governance influencing the type chosen. Our fundamental argument is that these two dimensions reinforce each other, especially in the two countries in question. The structures that have been developed in Norway, for example, are an apt match for the particular cultural setting within which they will function, as are those of the United States. Indeed, we argue that the structures overly reflect the culture, and perhaps do not enough to counter the decentralizing and antistate aspects of the culture.

Alternative Conceptions of Governance

Introduction

We will now proceed to briefly lay out three alternative conceptions of governance. These range from relatively straight-forward instrumental conceptions, through institutional conceptions, to individualistic and more philosophical examinations of the self-regulatory capacity of societies and the potential irrelevance of governance. In each section we will compare Norway and the United States in terms of how well they meet the criteria of governance that are implicit or explicit in the conceptualization, and with that to some extent also evaluate the adequacy of the concept of governance for use in comparative politics.

Instrumental Conceptions of Governance

We can begin with an instrumentally rational conception of governance, in which perspective the capacity of the political systems to establish coherent goals and then to devise means to reach those goals would be equated with governance (Peters 1997). Further, we might be concerned with the capacity of government to act swiftly and decisively and to be able to sustain its commitment to a set of broad policy goals. In this view, government is an "issue machine" (Braybrooke 1964), taking inputs from society in the form of issues and problems, processing them, producing policy goals as the output, and then finding ways to make those goals into a form of actuality for society.

Such an instrumental approach would assume that a reasonably clear and agreed-upon set of goals could be established for the entire society, preferably through some sort of democratic political process. With the goals established in the open forum, the process of governance would then become that of finding the instrumental means to achieve the goals, usually through a less manifestly democratic process. That subsequent implementation process need not, however, be dominated by the public sector itself; governments may "steer" rather than "row" (Osborne and Gaebler 1992) and use social actors as the principal sources of implementation activity.

Using this instrumental conception of governance, neither Norway nor the United States performs particularly well. In both countries, decisions tend to be made relatively slowly, and policies require a great deal of bargaining before being accepted for implementation. Despite the apparent strong *etatiste* tradition in Norway, there are a number of constraints on swift, decisive action; more predictably, government in the United States usually proceeds at an extremely deliberate pace. There are, however, significant differences in why governance does not proceed easily in the two countries.

The structural causes for the relative slowness and apparent indecisiveness of American government are readily apparent. The institutions of government were designed *not* to move quickly and to forestall action as long as possible. Within the federal government itself the separation of powers requires that the executive and legislature agree, and that the courts also find their decisions constitutional. The era of divided government currently being "enjoyed" in the United States means that those problems of governance have been exacerbated (Sundquist 1992; but see Jones [1995] and Mayhem [1991]), especially given the ideological character of at least one of the parties at this particular time.

Further, the current, crazy quilt pattern of intergovernmental relations means that attempting to produce change in any policy area may require getting literally thousands of separate governments "on board." The reverse side of the importance of federalism is that the states often can address problems that the federal government cannot. This can be seen very clearly with health-care reform, in which several states (such as Hawaii and Tennessee) have made significant advances in providing medical care for all their citizens (Leichter 1992), even though similar provisions have floundered at the federal level.

In addition to the structural problems inherent in divided government and federalism, there is internal fragmentation of public administration itself that makes American government less likely to be able to govern effectively. The demography and structure of bureaucracy tend to separate agencies from control by cabinet secretaries and the president and to produce substantial policy incoherence (Peters 1994). Civil servants tend to be recruited according to particular expertise and education, and to spend most of their careers within a single agency or department. Further, the ties of the agencies to equally specialized congressional committees tend to lessen the control capacity of other executive-department officials. Again, we encounter the familiar problem of government against subgovernment, with the subgovernments tending to win the battles.

Paradoxically, American government appears to be capable at once of too many decisions and too few decisions. The system has little difficulty in grinding out large numbers of detailed and segmented decisions, often decisions that benefit particular client groups.[2] Within each of the "silos" of policy the system appears to work well. On the other hand, the system has great difficulty in making comprehensive and coordinate decisions that operate across a range of interests and policy areas. Forming coalitions of this type are exceedingly difficult, so Congress tends to resort to vague laws, or no laws, leaving the problems of actually making policy to the numerous administrative agencies and state and local governments that must be involved in implementation.

The problems of governance in Norway, seen from the instrumental perspective, are really very different. The norms of the parliamentary system mean that even when there is a minority government (Strom 1990), the government is able to legislate when it wants or needs to do so. The form of parliamentary government practiced in Norway is rather different from the more adversarial forms found in many Westminster systems. The norms are less those of conflict and opposition and more of cooperation and working together to govern. Further, the cross-cutting nature of the issues and coalitions tends to minimize disruptive conflicts. Any parliamentary system provides a sitting government with a strong majority and thus the capacity to push through legislation, but the Norwegian model and to some extent those of other Scandinavian countries (Arter 1984) permit effective governance even in the face of ambiguous or nonexistent majorities.

Finally, the Norwegian civil service itself is more integrated so that there are fewer coordination and coherence problems than in American government. Demographically, the civil service and political elites tend to have relatively similar backgrounds, to have worked together for long periods of time, and to share many political ideals. Unlike the civil-service systems of many other industrialized democracies, these civil servants still work in a common "village" (Peters 1986). In such a structural setting, which provides effective government, from the instrumental perspective it appears to be a relatively easy task.

The barriers to rapid policy movements in Norway arise from the strong commitments to public participation that pervade the political system (Olsen

1983). In the Norwegian case many, if not most, of the potential barriers to effective decisions occur prior to the decision coming before the Storting for determination, and to some extent also occur at the implementation stage when affected interests have an ability to help shape the true meaning of the policy. The prior negotiation, however, tends to lessen the implementation debates (unlike in the United States) and make the policy as adopted by the legislature likely to be the policy that is implemented.

Interpretatively, both political systems visualize the barriers to policy action they present as an essential component of democracy. For the United States, the ingrained ideas of democracy require a concurrence of institutions so that ill-considered actions have little opportunity of becoming law. For Norway, democracy means the right of any and all affected interests to have a say about the policy, and potentially to block action if there is insufficient agreement across society. If nothing else, the Norwegian system can generate amendment of policies and reconciliation among the actors, something that parliamentary systems, it is argued, are not usually good at given their presumed capacity to bully policies through, if necessary (Weaver and Rockman 1993).

Institutional Conceptions of Governance

The instrumental conception of governance can be contrasted with a more institutionalist–cultural conception that places the emphasis on the legitimation of policy choices and the maintenance of norms rather than on goal attainment (March and Olsen 1989). An important implicit assumption of the institutionalists is that without the maintenance of norms and the attachment of legitimacy to a policy, the probabilities of successful goal attainment actually are very low. In this view, building the long-term normative basis for policy success is at least as important as winning or losing on the particular policy issue. Legitimacy arises from the long-term integration of citizens into the political process and a process of learning, rather than necessarily from winning or losing on every issue. Members of an effective democracy understand that they will not win every battle, but consider the capacity to engage in the policy debates essential to their well being.

Governance thus can be considered as the capacity of a governing system to minimize internal conflicts and organize the system to mediate conflicts in a manner that will permit that system to persist. In that conception, governance would be more closely associated with the task of legitimation than it is with specific goal attainment. Actually achieving anything through the political process may be secondary to simply including all the affected parties in the decision process and developing an algorithm for relating different policy preferences to more operational goals for the entire system. In the first approach to governance, we have assumed a coherent set of goals emerging from the process, while in this instance we should expect a great deal more variation.

In this case, if we contrast Norway and the United States we find somewhat different patterns of governance. Governing in the United States first requires a major act of legitimation for almost any form of activity. The skepticism about legitimacy of public acts is both constitutional and cultural. Although the bounds of the Constitution have been stretched substantially over time, any policy advocate must still locate a constitutional peg on which to hang the policy, with the courts serving as the judge of how well the justification and the policy match. For example, just days after the disastrous Oklahoma City bombing, the Supreme Court (*United States* v. *Lopez*) threw out the Gun-Free Schools Act of 1990 because the constitutional justification (the commerce clause) was inadequate for the particular piece of legislation. The public appeared to support this piece of legislation, but yet there was not a constitutional base and it was declared null.

There is also a cultural barrier to developing legitimate public responses to public problems. The American political psyche has perhaps even more limitations on government action than does the constitution. Again, the Oklahoma City bombing and its apparent "justification" represent the extreme limits of popular acceptance of public action, but most Americans have a narrower conception of the public than do most Europeans. Governing then requires determining just how far it is politically feasible to go in advocating new programs.

Once that first major hurdle of legitimation is passed, government in the United States is essentially an adversarial system of governing, with legitimation of policy eventually coming through the conflict among numerous interests and numerous institutions. As noted above, that conflict is in itself an important part of the legitimation process; if a policy cannot survive that trial then, it is thought, the country would probably be better off without it. The idea of consensual governance does not fit very well with the pluralistic notions that undergird most of the thinking about, and practice of, American politics. Consensus might quickly be seen as something like domination of the independent-minded American citizen by government rather than the opposite, which appears to be the real goal of the system.

As might be expected, governance problems in Norway are almost exactly the opposite as those encountered in the United States. There is little doubt that government has a very wide latitude for legitimate action in Norway, and the burden of proof appears to rest upon those who might want to restrict the level of state action rather than on those who want to increase it. There is definitely a tradition of individual liberty in Norway, but those norms appear to be tempered by an equally strong commitment to collective goals and collective goal attainment—especially given the enduring power of the Labor Party in government. Thus, the normative element of institutional governing appears relatively straightforward, and government has the latitude to act so long as its actions tend to be in the direction of "justice," conceptualized almost in a Rawlsian format of promoting the interests of the weakest elements of society (Rawls 1971).

Likewise, the formal institutional structure tends to make governing, or at least making collective decisions, relatively simple in Norway. The political system is unified at the national level, and unitary. Likewise, Norwegian national government is parliamentary, and is at that a very consensual and cooperative version of parliamentary government. The small size of the country and the relatively common elite backgrounds further reduce conflicts and enhance understandings within government. In other words, the system is well designed to produce strong and effective government, albeit within accepted and agreed-upon boundaries of legitimate action.

The paradox is that the Norwegian government appears better designed to deal with the somewhat fragmented and disorganized character of American society than with the more orderly Norwegian society. The various mechanisms for consultation available for government agencies provide a way of gauging organized public opinion and perhaps averting conflicts later in the policy process. In addition, the parliamentary system enables government to act more decisively than the slow and fragmented American political process. To an American, the Norwegian environment appears—if anything—too easy for government to govern.

Individualistic Conceptions of Governance

Finally, an economic liberal (see Buchanan 1992) might think that effective governance is only the establishment of the conditions within which individuals or interest organizations can achieve their own ends, almost without regard to the collective outcomes. This more atomistic conception of governing assigns a very limited role to the public sector and assumes that any form of coordination that is needed among individuals can be achieved through voluntary cooperation, rather than the coercion that would come through the public sector. Even when government must be involved, there is still room for complementary private action—for example, co-production of police services through devices such as "neighborhood watch."

One does not have to be an economic neoliberal, however, in order to consider how governance might be attained along this more individualistic format and through less-direct means. Indeed, a good deal of discussion on the political left concerning enhanced participation and debureaucratization has tended to push governance toward a more individualized, "empowerment" model. This conceptualization is in marked contrast to the more centralized conceptions with which we usually associate the term *governance*. In this decentralized conception of government, the public sector does not even "steer" (much less "row"), but rather establishes the normative and institutional framework within which other actors would be expected to steer and row (Veld 1992).

In extreme versions, it is argued that attempts at providing more centralized versions of governance are considered foolish and counterproductive, given the self-organizing capacity of systems and the resistance of individuals to many forms of external control. In such a view, a highly directive governance system may be effective for a short period of time but inevitably will run afoul of individual desires and social capacities for the evasion of rules. Governments would only be deceiving themselves if they believed that they had any significant control over the society.

Interestingly, from the perspective of the sociology of knowledge, the individualistic conception of governance is in some ways less developed in the United States than in other parts of the world. This is certainly not true for the libertarian, free-market conception, but it is for the other forms of self-organization. In Europe, ideas about self-organization and networks tend to be stronger (Rhodes 1997), and there is a sense that governance can be produced even without formal government action. For Americans, the constitutionalism of the political culture and the attendant emphasis on legality (procedural due process, for example) make attempts at supplying governance outside the system appear illegitimate. The foremost intellectual opponent of the use of non-governmental position is Theodore Lowi (1973).

As might be expected, individualistic conceptions of governing have substantially less resonance in Norway. This society and policy are much more receptive to collective goal-setting and policy-determination than in the United States, or indeed most other contemporary political systems. That said, Norway has not, however, been entirely immune to the spread of the more individualistic and antigovernmental ideas of the past two decades. Norway has its own homegrown antigovernment party (the Progress Party)—although modeled to some degree on the Danish example—and has engaged in, or at least considered, many of the same administrative reforms that other industrialized democracies have implemented. The spread of "ideas in good currency" makes it difficult for even the most committed collectivists to disregard possibilities about restructuring governance.

For Norway, the more individualistic forms of governance appear to imply more governance by lower levels of the public sector rather than devolution of power to the public as individuals. There has been a long and active debate in Norway over the idea of "free communes" and the capacity of local governments to make more of their own decisions with the control of Oslo. While these policy changes could hardly be said to be the equivalent of the more radical antigovernmental sentiments of the United States or some of the self-organizing logic of other European countries, they still do represent a departure from what has become the traditional mode of governance in Norway.

The communitarian movement (Etzioni 1993) is of substantial interest and importance at this point in the analysis. This growing political movement has argued that debureaucratization of public functions is crucial for maintaining both effectiveness and legitimacy for the public sector. This movement can be interpreted, however, as being on the political right or the political left.[3] On the one hand, this can be seen as a collectivist form of governance in which the community decides, and even implements, public policy. Individuals are important in this sense primarily as parts of the community. On the other hand, communitarianism can be seen as a highly individualistic way of keeping government, and especially central governments, out of people's everyday lives. In this view of political life, people through voluntary action will supply everything that is not absolutely beyond their capacity and hence would require government, including some services such as education that have become commonly thought of as public. Advocates of this view tend to regard few public functions as beyond the capacity of self-organized groups.

SUMMARY

There is probably some validity to each of the above views of governance. What is perhaps especially important for our comparative analysis is the extent to which governance is a socially constructed term. We as social scientists may want to impose a common definition on all governments, and to some extent we will do just that. In reality, however, each country will develop its own operating norms about governance and may tend to evaluate its own performance according to those standards.

Governance then, like many other political phenomena, is something of a social construction of reality (Schon and Rein 1994). The socially constructed nature of governance can be seen in the way in which the question of collective steering and conflict management are dealt with in the United States and Norway. In particular, we will be arguing that governance in the United States is very often about the symbolic aspects of governing, rather than about collective goal attainment. The complexity of the social and economic system, and the mirroring of that complexity in the governance apparatus, makes actual goal attainment more difficult to achieve. Further, the level of social fragmentation and pluralism that exists requires a great deal of collective myth creation and an investment in the mobilization of support. Therefore American politics talks a great deal about governance, but actually can govern very little. Even the meaning of governance in the United States is different from those that might be in other societies, even among other industrialized democracies. Given the magnitude of the tasks and the generally hostile environment for the public sector, governance tends to be more about influencing the environment of action than it is about influencing actions directly (Lowi 1973). This is especially true of *effective* governance in the United States; to get things done in the public sector often means going outside its formal structures. While many analysts such as Lowi may bewail the loss of public control and the absence of clear legal standards for behavior implied in that form of governance, the type of governance that is supplied may be the only type that is really possible under the circumstances.

Another important aspect of the social construction of governing in the United States is that governing tends to be conceived of in sectors rather than as a coherent and integrated activity. This perception is hardly confined to the United States (Muller 1985; Richardson and Jordan 1994), but it tends to reach its apotheosis there. Governing in the United States is very much "government against subgovernment" (Rose 1980) rather than the generation and imposition of common-goal statements for society. This pattern explains in part some of the policy problems of the United States (e.g., the continuing budgetary deficit). Thinking about the deficit requires substantial weighing of alternatives and trade-offs rather than promoting and protecting specific interests, and American politics is much better at the latter than the former.

The Norwegian conception of governance, on the other hand, is more precise and potentially also more instrumental in its focus. Governance is perhaps discussed less but done more; this is the Reebok approach to governing—"just do it." Given that Norwegian society is better integrated and also that social goals tend to be more uniform across the society, there is an enhanced capacity to get on with the tasks of governing. Because of the broad agreement on goals, if not on the relationship of every policy project to those goals, there will need to be less argument about first principles. Further, the greater hierarchical integration of government and of the social organizations that undergird that government will permit more governance "from the top-down" than might be effective in the United States.

The characterization of a direct style of governance in the Norwegian case must be balanced by remembering the strong role of participation and the imperatives of legitimation in the system. The governing system permits a variety of inputs from the society at the formulation and at the implementation stages, not to mention the more indirect participation through the legislative process. The cultural demands for broad and effective participation mean that policymaking usually does not proceed quickly, despite the apparent capacity to do so. This also means, however, that once government does decide to act, it can enforce its decisions rather easily, given the degree of consensus that will have been created already.

NOTES

1. Etymologically, the root of *governance* derives from words meaning "steering" or "controlling."

2. These are usually thought of as being privileged groups such as business. While it is certainly true that there are a number of decisions favorable to business, the social-service departments also churn out a large number of decisions favorable to their clients, and EPA and OSHA produce a large volume of decisions that are repugnant to business.

3. Most communitarians, however, would tend to reject these traditional labels. They would consider that both right and left in conventional political debate tend to miss out on the opportunities for cooperative actions, albeit perhaps for different reasons.

References

Aaron, H. J. 1994. "The Balanced Budget Blunder." *Brookings Review* (Spring): 41.

Aberbach, J. D., R. D. Putnam, and B. A. Rockman, eds. 1981. *Bureaucrats and Politicians in Western Democracies*. Cambridge, MA: Harvard University Press.

Aberbach, J. D., and B. A. Rockman. 1976. "Clashing Beliefs within the Executive Branch: The Nixon Administration and the Bureaucracy." *American Political Science Review* 70: 456–68.

ACIR. 1994. *Changing Public Attitudes on Government and Taxes*. Washington, DC: Advisory Commission on Intergovernment Relations.

Adler, E. 1992. "The Emergence of Cooperation: National Epistemic Communities and International Evolution of the Idea of Nuclear Arms Control." *International Organization* 46: 101–46.

Allard, C. 1990. *Command, Control and the Common Defense*. New Haven, CT: Yale University Press.

Allison, G. T. 1971. *Essence of Decision*. Boston: Little, Brown.

———. 1987. "Public and Private Management: Are They Fundamentally Alike in All Unimportant Respects?" In *Classics of Public Administration*, edited by J. A. Shafritz and A. C. Hyde. Homewood, IL: Dorsey.

Anagnoson, J. T. 1989. "Administrative Goals, Environments, and Strategies." In *Home Style and the Washington Work: Studies of Congressional Politics*, edited by M. P. Fiorina and D. W. Rohde. Ann Arbor: University of Michigan Press.

Arnold, P. E. 1986. *Making the Managerial Presidency: Comprehensive Reorganization Planning 1905–1980*. Princeton, NJ: Princeton University Press.

Arter, D. 1984. *The Nordic Parliaments: A Comparative Analysis*. New York: St. Martin's Press.

Bakema, W. E. 1991. "The Ministerial Carreer." In *The Profession of Government Minister in Western Europe*, edited by J. Blondel and J. L. Thiebault. Basingstoke, UK: Macmillan.

Bartolini, S. 1993. "Of Time and Comparative Research." *Journal of Theoretical Politics* 5: 131–67.

Barzelay, M. 1993. *Breaking through Bureaucracy*. Berkeley: University of California Press.

Beard, C. A. 1935. *An Economic Interpretation of the Constitution of the United States.* New York: Macmillan.
Benda, P., and C. H. Levine. 1986. "Reagan and the Bureaucracy: The Bequest, the Promise and the Legacy." In *The Reagan Legacy*, edited by C. O. Jones. Chatham, NJ: Chatham House.
Benum, E. 1979. *Sentraladministrasjonens historie.* Bind 2. 1845–84. Oslo: Norwegian University Press.
Berggrav, D. 1985. "Regjeringen." In *Storting og regjering 1945–1985. Institusjoner og rekruttering*, edited by T. Nordby. Oslo: Kunnskapsforlaget.
———. 1994. *Slik styres Norge. Kongen, regjeringen ogStortinget i norsk statsliv.* Oslo: Schibsted.
Berman, L. 1987. *The New American Presidency.* Boston: Little, Brown.
Blondel, J. 1985. *Government Ministers in the Contemporary World.* London: Sage.
Blondel, J., and J.-L. Thiebault. 1991. *The Profession of Government Minister in Western Europe.* Basingstoke, UK: Macmillan.
Bratbak, B., and J. P. Olsen. 1980. "Departement og opinion: Tilbakeføring av informasjon om virkninger av offentlige tiltak." In *Meninger og makt*, edited by J. P. Olsen. Bergen: Norwegian University Press.
Braybrooke, D. 1964. *Traffic Congestion Goes through the Issue Machine.* London: Routledge and Kegan Paul.
Bryner, G. 1986. *Bureaucratic Discretion.* New York: Pergammon.
Buchanan, J. M., and G. Tullock. 1962. *The Calculus of Consent.* Ann Arbor: University of Michigan Press.
Burke, J. P. 1992. *The Institutional Presidency.* Baltimore: Johns Hopkins University Press.
Burns, J. 1956. *Roosevelt: The Lion and the Fox.* New York: Harcourt, Brace.
Caldera, G., and J. R. Wright. 1990. "Amici Curiae Before the Supreme Court: Who Participates, When and How Much." *Journal of Politics* 52: 782–806.
Calvert, R., M. McCubbins, and B. Weingast. 1989. "A Theory of Political Control of Agency Discretion." *American Journal of Political Science* 33: 588–611.
Campbell, C., and B. G. Peters. 1988. "Introduction." In *Organizing Governance: Governing Organizations*, edited by C. Campbell and B. G. Peters. Pittsburgh: University of Pittsburgh Press.
Caro, R. A. 1982. *The Years of Lyndon Johnson.* New York: Knopf.
Castles, F. G. 1993. *Families of Nations: Patterns of Public Policy in Western Democracies.* Aldershot, UK: Dartmouth.
Christensen, T. 1987. "How to Succeed in Reorganizing: The Case of the Norwegian Health Administration." *Scandinavian Political Studies* 10: 61–77.
———. 1989. "Turnover Intentions in Public Bureaucracies: Explanations and Implications." Manuscript. Department of Political Science, University of Oslo.
———. 1991a. "Bureaucratic Roles: Political Loyalty and Professional Autonomy." *Scandinavian Political Studies* 14: 303–20.
———. 1991b. *Virsomhetsplanlegging-instrumentell problemløsning eller myteskaping?* Oslo: TANO.
———. 1994a. *Politisk styring og faglig uavhengighet. Reorganisering av den sentrale helseforvaltningen.* Oslo: TANO.

———. 1994b. "Administrative Reform: The Comparative Potential of Instrumental and Institutional Theory." Manuscript. Department of Political Science, University of Oslo.

———. 1996. "Adapting to Processes of Europeanisation: A Study of the Norwegian Ministry of Foreign Affairs." ARENA Report No.2.

———. 1997a. "Utviklingen av direktoratene aktører, tenkning og organisasjonsformer." In *Forvaltningskunnskap*, edited by T. Christensen and M. Egeberg. Oslo: Aschehoug Tano.

———. 1997b. "Structure and Culture Reinforce: The Development and Current Features of the Norwegian Civil Service System." Paper presented at the Civil Service in Comparative Perspective Conference, Indiana University, April.

Christensen, T., and M. Egeberg. 1979. "Organized Group-Government Relations in Norway: On the Structured Selection of Participants, Problems, Solutions and Choice Opportunities." *Scandinavian Political Studies* 3(2): 239–60.

———. 1997a. "Sentraladministrasjonenen oversikt over trekk ved departement og direktorat." In *Forvaltningskunnskap*, edited by T. Christensen and M. Egeberg. Oslo: Aschehoug Tano.

———. 1997b. "Noen trekk ved forholdet mellom organisasjonene og den offentlige forvaltningen." In *Forvaltningskunnskap*, edited by T. Christensen and M. Egeberg. Oslo: Aschehoug Tano.

Christensen, T., and P. Laegreid. 1996. "Administrative Policy in Norway: Towards New Public Management?" LOS Report No. 47, University of Bergen.

Cohen, J. E. 1988. *The Politics of the U.S. Cabinet*. Pittsburgh: University of Pittsburgh Press.

Comarow, M. 1981. "The War on the Civil Servants." *Bureaucrat* 10: 8–9.

Cotta, M. 1991. "Conclusion." In *The Profession of Government Minister in Western Europe*, edited by J. Blondel and J.-L.Thiebault. Basingstoke, UK: Macmillan.

Cox, G. W., and S. Kernell. 1991. "Introduction: Governing a Divided Era." In *The Politics of Divided Government*, edited by G. W. Cox and S. Kernell. Boulder, CO: Westview Press.

Cronin, T. E. 1975. *The State of the Presidency*. Boston: Little, Brown.

Dahl, R. A., and C. E. Lindblom. 1953. *Politics, Economics and Welfare*. New York: Harpers.

Danielsen, R. 1964. *Det Norske Storting gjennom 150 år*. Bind II. *Tidsrommet 1870–1908*. Oslo: Gyldendal.

Davidson, R. H. 1988. "'Invitation to Struggle': An Overview of Legislative Executive Relations." In *Congress and the Presidency: Invitation to Struggle. The Annals of the American Academy of Political and Social Science*, edited by R. H. Davidson. Newbury Park, CA: Sage.

———. 1992. *The Postreform Congress*. New York: St. Martin's Press.

Debes, J. 1950. *Det Norske Statsråd 1814–1949*. Oslo: Cammermeyers Boghandel.

Derthick, M. 1990. *Agency Under Stress: The Social Security Administration in American Government*. Washington, DC: Brookings Institution.

Destler, I. M. 1985. "Executive–Congressional Conflict in Foreign Policy: Explaining It, Coping with It." In *Congress Reconsidered*, 3d ed., edited by L. C. Dodd and B. I. Oppenheimer. Washington, DC: CQ Press.

de Winter, L. 1991. "Parliamentary and Party Pathways to the Cabinet." In *The Profession of Government Minister in Western Europe*, edited by J. Blondel and J.-L. Thiebault. Basingstoke, UK: Macmillan.
Dexter, L. A. 1990. "Intra-Agency Politics: Conflict and Contravention in Administrative Entities." *Journal of Theoretical Politics* 2: 151–72.
DiIulio, J. 1994. *Deregulating Government.* Washington, DC: Brookings Institution.
Dodd, L. C. 1993. "Congress and the Politics of Renewal: Redressing the Crisis of Legitimation." In *Congress Reconsidered*, 5th ed., edited by L. C. Dodd and B. I. Oppenheimer. Washington, DC: CQ Press.
Dodd, L. C., and B. I. Oppenheimer. 1993. "Maintaining Order in the House: The Struggle for Institutional Equilibrium." In *Congress Reconsidered*, 5th ed., edited by L. C. Dodd and B. I. Oppenheimer. Washington, DC: CQ Press.
Douglas, M. 1986. *How Institutions Think.* Syracuse, NY: Syracuse University Press.
———. 1990. "Risk as a Forensic Resource." *Daedalus* 119(4): 1–16.
Easton, D. 1965. *A Systems Analysis of Political Life.* New York: Wiley.
Egeberg, M. 1981. *Stat og organisasjoner.* Oslo: Norwegian University Press.
———. 1984. *Organisasjonsutforming i o ffentlig virksomhet.* Oslo: TANO.
———. 1987. "Designing public organizations." In *Managing Public Organizations,* edited by J. Kooiman and K. A. Eliassen. London: Sage.
———. 1989a. "Effekter av organisasjonsendring i forvaltningen." In *Institusjonspolitikk og forvaltningsutvikling. Bidrag til en anvendt statsvitenskap*, edited by M. Egeberg. Oslo: TANO.
———. 1989b. "Om å organisere konkurrerende beslutningsprinsipper inn imyndighetsstrukturer." In *Institusjonspolitikk og forvaltningsutvikling. Bidrag til en anvendt statsvitenskap*, edited by M. Egeberg. Oslo: TANO.
———. 1997. "Verdier i statsstyre." In *Forvaltnings-kunnskap*, edited by T. Christensen and M. Egeberg. Oslo: Aschehoug Tano.
Eisner, M. A. 1991. *Antitrust and the Triumph of Economics.* Chapel Hill: University of North Carolina Press.
———. 1993. "Bureaucratic Professionalism and the Limits of Political Control Thesis: The Case of the Federal Trade Commission." *Governance* 6: 127–53.
Eliassen, K. A. 1985. "Rekrutteringen til Stortinget og regjering 1945–84." In *Storting og regjering 1945–85. Institusjoner rekruttering*, edited by T. Norby. Oslo: Kunnskapsforlaget.
Elkins, D. J., and R. E. B. Simeon. 1979. "A Cause in Search of an Effect; Or What Does Political Culture Explain." *Comparative Politics* 11: 117–46.
Eriksen, S. 1988a. "Norway: Ministerial Autonomy and Collective Responsibility." In *Cabinets in Western Europe*, edited by J. Blondel and F. Muller-Rommel. Basingstoke, UK: Macmillan.
———. 1988b. *Herskap og tjenere Om samarbeidet mellom politikere og tjenestemenn i departementene.* Oslo: TANO.
Etzioni, A. 1975. *A Comparative Analysis of Complex Organizations*, rev. ed. New York: Free Press.
———. 1993. *The Spirit of Community: Rights, Responsibilities and the Communitarian Agenda.* New York: Crown.

Evans, P. 1995. *Embedded Autonomy: States and Industrial Transformation*. Princeton, NJ: Princeton University Press.

Evans, P., D. Rueschmeyer, and T. Skocpol, eds. 1985. Bringing the State Back In. Cambridge: Cambridge University Press.

Farsund, A. A., and U. Sverdrup. 1994. "Norsk forvaltning i EØP prosessen. Noen empiriske resultat." *Nordisk Administrativt Tidsskrift* 75(1): 48–67.

Fenno, R. F. 1978. *Home Style: House Members in Their Districts*. Boston: Little, Brown.

———. 1987. "Congressmen and Committees: A Comparative Analysis." In *Congress: Structure and Policy*, edited by M. D. McCubbins and T. Sullivan. Cambridge: Cambridge University Press.

Fiorina, M. P. 1987. "The Case of the Vanishing Marginals: The Bureaucracy Did It." In *Congress: Structure and Policy*, edited by M. D. McCubbins and T. Sullivan. Cambridge: Cambridge University Press.

———. 1989. *Congress: Keystone of the Washington Establishment*. New Haven, CT: Yale University Press.

———. 1991. "Coalition Governments, Divided Governments and Electoral Theory." *Governance* 4: 236–49.

———. 1992. "An Era of Divided Government?" *Political Science Quarterly* 107: 387–410.

Fiorina, M. P., and D. W. Rohde. 1989. "Richard Fenno's Research Agenda and the Study of Congress." In *Home Style and the Washington Work: Studies of Congressional Politics*, edited by M. P. Fiorina and D. W. Rohde. Ann Arbor: University of Michigan Press.

Fisher, L. 1991. *Constitutional Conflicts between Congress and the President*, 3d ed. Lawrence: University of Kansas Press.

Freedman, J. O. 1980. *Crisis and Legitimacy*. Cambridge: Cambridge University Press.

Freeman, G. 1985. "National Styles and Policy Sectors: Explaining Structured Variation." *Journal of Public Policy* 5: 467–96.

Frendreis, J. P. 1983. "Explanation of Variation and Detection of Covariation: The Purpose and Logic of Comparative Analysis." *Comparative Political Studies* 16: 255–73.

Friedson, E. 1986. *Professional Powers*. Chicago: University of Chicago Press.

Fukuyama, F. 1995. *Trust: Social Virtues and the Creation of Prosperity*. New York: Free Press.

Gallup Monthly Report. 1996. "Confidence in Institutions" (June).

Goodsell, C. 1995. *The Case for Bureaucracy*, 3rd. ed. Chatham, NJ: Chatham House.

Gore Commission/National Performance Review. 1993. *Creating a Government That Works Better and Costs Less*. Washington, DC: Government Printing Office.

Gormley, W. T. 1989. *Taming the Bureaucracy: Muscles, Prayers, and Other Strategies*. Princeton, NJ: Princeton University Press.

Gulick, L. 1937. "Notes on the Theory of Organizations: With Special Reference to Government." In *Papers on the Science of Administration*, edited by L. Gulick and L. Urwick. Chicago: A. M. Kelley.

Gulick, L., and L. Urwick, eds. 1937. *Papers on the Science of Administration*. Chicago: A. M. Kelley.

Haas, E. B. 1990. *When Knowledge Is Power: Three Models of Change in International Organizations*. Berkeley: University of California Press.

Hacker, A. 1992. *Two Nations: Black and White, Separate, Hostile, Unequal*. New York: Scribners.

Hall, P. A. 1989. *The Power of Economic Ideas*. Princeton, NJ: Princeton University Press.

Hall, R. L. 1993. "Participation, Abdication, and Representation in Congressional Committees." In *Congress Reconsidered*, 5th ed., edited by L. C. Dodd and B. I. Oppenheimer. Washington, DC: CQ Press.

Hanuschek, E. A. 1986. "Formula Budgeting: The Economics and Politics of Fiscal Policy Under Rules." *Journal of Policy Analysis and Management* 6: 3–19.

Hargrove, E. C. 1994. *Prisoners of Myth: The Leadership of the Tennessee Valley Authority, 1933–90*. Princeton, NJ: Princeton University Press.

Hart, J. 1987. *The Presidental Branch*. Chatham, NJ: Chatham House.

———. 1994. *The Presidental Branch*, 2d ed. Chatham, NJ: Chatham House.

Heclo, H. 1977. *A Government of Strangers*. Washington, DC: Brookings Institution.

Heidar, K. 1983. *Norske politiske fakta 1884–1982*. Oslo: Norwegian University Press.

———. 1988. *Partidemokrati på prøve. Norske partieliter i demokratisk perspektiv*. Oslo: Norwegian University Press.

Heidar, K., and E. Berntzen. 1993. *Vesteuropeisk politikk. Partier, regjeringsmakt, styreform*. Oslo: Norwegian University Press.

Hellevik, O. 1969. *Stortinget—en sosial elite?* Oslo: Pax.

Hernes, G. 1971. "Interest, Influence and Cooperation: A Study of the Norwegian Parliament." Unpublished Ph.D. thesis, Johns Hopkins University. Baltimore, MD.

———. 1977. "Interest and the Structure of Influence." In *The History of Parliament*, edited by W. O. Aydelotte. Princeton, NJ: Princeton University Press.

Hernes, G., and K. Nergaard. 1990. *Oss i mellom. Konstitusjonelle former og uformelle kontakter Storting Regjering*. Oslo: FAFO.

Hibbing, J. R. 1993. "Careerism in Congress: For Better or for Worse?" In *Congress Reconsidered*, 5th ed., edited by L. C. Dodd and B. I. Oppenheimer. Washington, DC: CQ Press.

Himmelberg, R. F. 1994. *Survival of Corporatism During the New Deal Era, 1933–1945*. New York: Garland.

Hoff, G. 1964. "Stortingets administrasjon, herunder Stortingets arkiv og bibliotek." In *Det Norske Storting gjennom 150 år*, edited by K. Kaartvedt, R. Danielsen, and T. Greve. Bind IV. Spesialartikler. Oslo: Gyldendal.

Hofstede, G. 1991. *Cultures and Organizations: The Software of the Mind*. New York: McGraw-Hill.

Hogwood, B. W., and B. G. Peters. 1983. *Policy Dynamics*. Brighton, UK: Wheatsheaf.

Hood, C., and M. Jackson. 1991. *Administrative Argument*. Aldershot, UK: Dartmouth.

Hoogenboom, A. 1968. *Outlawing the Spoils: A History of the Civil Service Reform Movement, 1865–1883*. Urbana: University of Illinois Press.

Hughes, J. R. T. 1991. *The Governmental Habit Redux: Economic Controls from Colonial Days to the Present*. Princeton, NJ: Princeton University Press.

Hult, K. 1987. *Agency Merger and Bureaucratic Redesign*. Pittsburgh: University of Pittsburgh Press.

Hult, K., and C. Walcott. 1990. *Governing Public Organizations: Politics, Structures, and Institutonal Design*. Pacific Grove, CA: Brooks/Cole Publishing.

Huntington, S. P. 1952. "The Marasmus of the ICC." *Yale Law Review* (April): 497–509.

Immergut, E. 1992. *Health Care Politics: Ideas and Institutions in Western Europe*. Cambridge: Cambridge University Press.

Ingraham, P. W., and C. Ban. 1984. *Legislating Bureaucratic Change: The Civil Service Reform Act of 1978*. Albany: State University Press of New York.

Ingraham, P. W., and D. H. Rosenbloom. 1992. *The Promise and Paradox of Civil Service Reform*. Pittsburgh: University of Pittsburgh Press.

Jacobsen, K. D. 1960. "Lojalitet, nøytralitet og faglig uavhengighet i sentraladministrasjonen." *Tidsskrift for Samfunnsforskning* 1: 231–48.

———. 1977. *Teknisk hjelp og politisk struktur*, 2d ed. Oslo: Norwegian University Press.

Jacobson, G. C. 1987. "Running Scared: Elections and Congressional Politics in the 1980s." In *Congress: Structure and Policy*, edited by M. D. McCubbins and T. Sullivan. Cambridge: Cambridge University Press.

Jenkins-Smith, H., and P. A. Sabatier. 1993. *Policy Change and Policy Learning*. Boulder, CO: Westview Press.

Johnson, R. N., and G. D. Libecap. 1994. *The Federal Civil Service System and the Problem of Bureaucracy*. Chicago: University of Chicago Press.

Jones, C. O. 1995. *Separate But Equal Branches: Congress and the Presidency*. Chatham, NJ: Chatham House.

Kaartvedt, A. 1964. *Det Norske Storting gjennom 150 år. Bind I. Fra Riks forsamlingen til 1869*. Oslo: Gyldendal.

Kaiser, F. M. 1988. "Congressional Oversight of the Presidency." In *Congress and the Presidency: Invitation to Struggle. The Annals of the American Academy of Political and Social Sciences*, edited by R. H. Davidson. Newbury Park, CA: Sage.

Kaplan, M. 1986. *The Great Society and Its Legacy*. Durham, NC: Duke University Press.

Kazin, M. 1995. *The Populist Persuasion: An American History*. New York: Basic Books.

Kelman, S. 1982. "Adversary and Cooperationist Institutions for Conflict Resolution in Public Policymaking." *Journal of Policy Analysis and Management* 11: 178–206.

Kerlinger, F. 1986. *Foundations of Behavioral Research*, 6th ed. Chicago: Rand McNally.

Kernaghan, H. 1992. "Empowerment and Public Administration: Revolutonary Advance or Passing Fancy?" *Canadian Public Administration* 35: 194–214.

Kernell, S. 1989. "The Evolution of the White House Staff." In *Can the Government Govern?*, edited by J. E. Chubb and P. E. Peterson. Washington, DC: Brookings Institution.

———. 1991. "Facing an Opposition Congress: The President's Strategic Circumstance." In *The Politics of Divided Government*, edited by G. W. Cox and S. Kernell. Boulder, CO: Westview Press.

Kerwin, M. 1994. *Rulemaking: How Government Agencies Write Law and Make Policy*. Washington, DC: CQ Press.

Kettl, D. F. 1984. "The Maturing of American Federalism." In *The Costs of Federalism: Essays In Honor of James W. Fesler*, edited by R. T. Golembiewski and A. Wildavsky. New Brunswick, NJ: Tranaction Books.

———. 1992. *Deficit Politics*. New York: Macmillan.

King, A. 1993. "Foundations of Power." In *Researching the Presidency: Vital Questions, New Approaches*, edited by G. E. Edwards III, J. H. Kessel, and B. A. Rockman. Pittsburgh: University of Pittsburgh Press.

King, D., and B. G. Peters. 1994. "The United States." In *Rewards at the Top*, edited by C. Hood and B. G. Peters. London: Sage.

Kooiman, J. 1993. *Modern Governance*. London: Sage.

Krasner, S. 1988. "Sovereignty: An Institutional Perspective." *Comparative Political Studies* 21(1): 66–94.

Kvavik, R. 1976. *Interest Groups in Norwegian Politics*. Oslo: Norwegian University Press.

Laegreid, P. 1989. *Rekrutteringspolitikk i sentraladministrasjonen*, paper no. 31. Bergen: Norwegian Research Centre in Organization and Management.

———. 1993a. *Rewards for High Public Office: The Case of Norway*, report no. 11. Department of Administration and Organization Theory, University of Bergen.

———. 1993b. *Salary Policy Reforms for High Public Office in Norway*, report no. 9320. Norwegian Research Centre in Organization and Management.

———. 1994. "Norway." In *Rewards at the Top*, edited by C. Hood and B. G. Peters. London: Sage.

Laegreid, P., and J. P. Olsen. 1978. *Byråkrati og beslutninger*. Bergen: Norwegian University Press.

Laegreid, P., and P. G. Roness. 1983. *Sentraladministrasjonen*. Oslo: Tiden Norsk Forlag.

Lafferty, W. 1981. *Participation and Democracy in Norway*. Oslo: Universitetsforlaget.

LaPalombara, J. 1968. "MacroTheories and Micro-Applications: A Widening Chasm." *Comparative Politics* 26: 461–77.

Laumann, E. O., and D. Knoke. 1987. *The Organizational State*. Madison: University of Wisconsin Press.

Laurin, U. 1986. *Paa Heder och Samvete*. Stockholm: Norstedts.

Leichter, H. M. 1992. *Health Reform in America: Innovations From the States*. Armonk, NY: M. E. Sharpe.

Levine, C. H., and R. S. Kleman. 1993. "The Quiet Crisis of the American Public Service." In *Agenda for Excellence: Public Service in America*, edited by P. W. Ingraham and D. F. Kettl. Chatham, NJ: Chatham House.

Light, P. C. 1993. *Monitoring Government: Inspectors General and the Search for Accountability in Government*. Washington, DC: Brookings Institution.

———. 1996. *Thickening Government: Federal Hierarchy and the Diffusion of Accountability*. Washington, DC: Brookings Institution.

Lijphart, A. 1971. "Comparative Politics and the Comparative Method." *American Political Science Review* 65: 682–93.

———. 1975. "The Comparable-Cases Strategy in Comparative Research." *Comparative Political Studies* 8: 158–77.

———. 1984. *Democracies: Patterns of Majoritarian and Concensus Government in Twenty-One Countries*. New Haven, CT: Yale University Press.

———. 1989. "Democratic Political Systems: Types, Cases, Caus[illegible] Consequences." *Journal of Theoretical Politics* (1): 33–48.

———. 1991. "Democratic Political Systems." In *Contemproary Political Systems: Class-*

ifications and Typologies, edited by A. Bebler and J. Seroka. Boulder, CO: Lynne Rienner.
———. 1977. *Politics and Markets: The World's Political Economic Systems*. New York: Basic Books.
Linz, J. J., and A. de Miguel. 1966. "Within-Nation Differences and Comparisons: The Eight Spains." In *Comparing Nations: The Use of Quantitative Data in Cross-National Research*, edited by R. L. Merritt and S. Rokkan. New Haven, CT: Yale University Press.
Listhaug, O., and A. H. Miller. 1985. "Public Support for Tax Evasion: Self Interest or Symbolic Politics." *European Journal for Political Research* 13: 265–82.
Lowi, T. J. 1973. *The End of Liberalism*. New York: Norton.
Mackenzie, G. C. 1987. *The In and Outers*. Baltimore: Johns Hopkins University Press.
Maidment, R., and A. McGrew. 1991. *The American Political Process*. London: Sage.
Malbin, M. J. 1980. *Unelected Representatives: Congressional Staff and the Future of Representative Government*. New York: Basic Books.
March, J. G., and J. P. Olsen. (1976). *Ambiguity and Choice in Organizations*. Oslo: Norwegian University Press.
———. 1983. "Organizing Political Life: What Administrative Reorganization Tells Us About Government." *American Political Science Review* 77: 281–97.
———. 1989. *Rediscovering Institutions: The Organizational Basis of Politics*. New York: Free Press.
———. 1995. *Democratic Governance*. New York: Free Press.
March, J. G., and H. Simon. 1958. *Organizations*. New York: Wiley.
Martens, H. 1979. *Centralforvaltning i seks vestlige lande*, 2d ed. Aarhus: Politica.
Mashaw, J. L. 1983. *Bureaucratic Justice*. New Haven, CT: Yale University Press.
Maurseth, P. 1979. *Sentraladministrasjonens historie 1814–1844*, bind 1. Oslo: Norwegian University Press.
Mayhew, D. R. 1989. "Does It Make a Difference Whether Party Control of the American National Government is Unified or Divided?" Paper presented at the Annual Meeting of the American Political Science Association, Atlanta, Ga.
———. 1991. *Divided We Govern: Party Control, Lawmaking, and Investigations, 1946–1990*. New Haven, CT: Yale University Press.
Mayntz, R., and F. W. Scharpf. 1975. *Policy-Making in the German Federal Bureaucracy*. New York: Elsevier.
McCubbins, M. D. 1991. "Government on Lay-Away: Federal Spending and Deficit Under Divided Party Control." In *The Politics of Divided Government*, edited by G. W. Cox and S. Kernell. Boulder, CO: Westview Press.
McCubbins, M. D., and T. Page. 1987. "A Theory of Congressional Delegation." In *Congress: Structure and Policy*, edited by M. D. McCubbins and T. Sullivan. Cambridge: Cambridge University Press.
McCubbins, M. D., and T. Schwartz. 1987. "Congressional Oversight Overlooked: Police Patrols Versus Fire Alarm." In *Congress: Structure and Policy*, edited by M. D. McCubbins and T. Sullivan. Cambridge: Cambridge University Press.
McCubbins, M. D., and T. Sullivan, eds. 1987. *Congress: Structure and Policy*. Cambridge: Cambridge University Press.

McGarrity, T. 1991. *Reinventing Rationality: Regulatory Analysis in the Federal Government.* Cambridge: Cambridge University Press.

Meyer, J., and B. Rowan. 1977. "Institutionalized Organizations. Formal Structure as Myth and Ceremony." American Journal of Sociology 83 (September): 340–63.

Meyer. 1985. *La politisation de l'administration publique.* Brussels: IIAS.

Mezey, M. L. 1985. "President and Congress: A Review Article." *Legislative Studies Quarterly* 10: 519–36.

———. 1991. "Congress Within the U.S. Presidential System." In *Divided Democracy: Cooperation and Conflict Between the President and Congress,* edited by J. A. Thurber. Washington, DC: CQ Press.

Mill, J. S. 1872. *A System of Logic,* 8th ed. London: Longmans, Green, Reader, and Dyer.

Moe, R. C. 1994. "The Reinventing Government Exercise: Misinterpreting the Problem Misjudging the Consequences." *Public Administration Review* 54: 212–34.

Moe, T. M. 1985. "The Politicized Presidency." In *The New Direction in American Politics,* edited by J. E. Chubb and P. E. Peterson. Washington, DC: Brookings Institution.

———. 1989. "The Politics of Bureaucratic Structure." In *Can the Government Govern?,* edited by J. E. Chubb and P. E. Peterson. Washington DC: Brookings Institution.

———. 1990. "Political Institutions: The Neglected Side of the Story." *Journal of Law, Economics, and Organization* 6(special issue): 213–53.

Morris, G., and S. Gates. 1986. *Inspectors General: Junkyard Dogs?* Washington, DC: Brookings Institution.

Morrow, W. L. 1987. "The Pluralist Legacy in the American Public Administration." In *A Centennial History of the American Administrative State,* edited by R. C. Chandler. New York: Free Press.

Moynihan, D. P. 1972. *Maximum Feasible Misunderstanding: Community Action in the War on Poverty.* New York: Free Press.

Muller, 1985. "Un schema d'analyse des politiques sectorielles." *Revue Francaise de Science Politique* 35: 165–89.

Nathan, R. P. 1975. *The Plot That Failed: Nixon and the Administrative Presidency.* New York: Wiley.

———. 1983. *The Administrative Presidency.* New York: Macmillan.

Neu, C. E. 1987. "The Rise of the National Security Bureaucracy." In *The New American State,* edited by L. Galambos. Baltimore: Johns Hopkins University Press.

Neustadt, R. 1960. *Presidential Power: The Politics of Leadership.* New York: Wiley.

Nordby, T., ed. 1993. *Arbeiderpartiet og planstyret 1945–65.* Oslo: Norwegian University Press.

North, D. C. 1991. *Institutions, Institutional Change and Economic Development.* Cambridge: Cambridge University Press.

Oleszek, W. J. 1991. "The Context of Congressional Policy Making." In *Divided Democracy. Cooperation and Conflict Between the President and Congress,* edited by J. A. Thurber. Washington, DC: CQ Press.

Olsen, J. P. 1983. *Organized Democracy.* Bergen: Norwegian University Press.

———. 1988a. *Statsstyre og institusjonsutforming.* Bergen: Norwegian University Press.

———. 1988b. "Administrative Reform and Theories of Organization." In *Organizing*

Governance: Governing Organizations, edited by C. Campbell and B. G. Peters. Pittsburgh: University of Pittsburgh Press.

———. 1995a. "Norway: Reluctant Reformer, Slow Learner, or Another Triumph for the Tortoise." In *Learning from Administrative Reform*, edited by J. P. Olsen and B. G. Peters. Oslo: Scandinavian University Press.

———. 1995b. "Norway: Reluctant Learner." In *Learning from Administrative Reform*, edited by J. P. Olsen and B. G. Peters. Oslo: Scandinavian University Press.

Olsen, J. P., and B. G. Peters 1996. *Lessons from Experience*. Oslo: Scandinavian University Press.

Ornstein, N. J., ed. 1982. *President and Congress: Assessing Reagan's First Year.* Washington, DC: American Enterprise Institute.

Ornstein, N. J., R. L. Peabody, and D. W. Rohde. 1993. "The U.S. Senate in an Era of Change." In *Congress Reconsidered*, 5th ed., edited by L. C. Dodd and B. I. Oppenheimer. Washington, DC: CQ Press.

Osborne, D., and T. Gaebler. 1992. *Reinventing Government: How the Entrepreneurial Spirit is Transforming the Public Sector.* Reading, MA: Addison-Wesley.

Ott, J. 1989. *The Organizational Culture Perspective.* Pacific Grove, CA: Brooks-Cole.

Page, E. C. 1992. *Political Authority and Bureaucratic Power: A Comparative Perspective*, 2d ed. New York: Harvester Wheatsheaf.

Pempel, T. J. 1990. *Uncommon Democracies. The One-Party Dominant Regimes.* Ithaca, NY: Cornell University Press.

Peters, B. G. 1985. "The United States: Absolute Change and Relative Stability." In *Public Employment in Western Countries*, edited by R. Rose et al. Cambridge: Cambridge University Press.

———. 1986. "Burning the Village: The Civil Service Under Reagan and Thatcher." *Parliamentary Affairs* 39: 79–97.

———. 1989. *The Politics of Bureaucracy*, 3d ed. New York: Longman.

———. 1991. *The Politics of Tax Policy*. Oxford: Blackwells.

———. 1994. *The Politics of Bureaucracy*, 4th ed. New York: Longman.

———. 1997. "Bringing the State Back in Again." Paper presented at the Conference on Governance, Loch Lomond, Scotland.

———. 1999. "The Institutions of Competition Policy in the United States." In *Competition Policy in Comparative Perspective*, edited by B. Doern and S. Wilks. Oxford: Oxford University Press.

Peters, B. G., and D. J. Savoie. 1994. "Reinventing Osborne and Gaebler: Lessons from the Gore Commission." *Canadian Public Administration.*

Peters, T. J., and H. Waterman. 1982. *In Search of Excellence*. New York: Harper & Row.

Pierre, J. 1995. "Citizen as Consumer: The Marketization of the State." In *Governance in a Changing Environment*, edited by D. J. Savoie and B. G. Peters. Montreal: McGill–Queens University Press.

Polsby, N. W. 1987. "The Institutionalization of the U.S. House of the Representatives." In *Congress: Structure and Policy*, edited by M. D. McCubbins and T. Sullivan. Cambridge: Cambridge University Press.

———. 1989. *Policy Innovation in the United States*. New Haven, CT: Yale University Press.

———. 1990. "Congress-Bashing for Beginners." *Public Interest* 100 (summer): 15–23.

Przeworski, A., and H. Teune. 1970. *The Logic of Comparative Social Inquiry*. New York: Wiley.

Public Perspective, The. 1994. Report on "Faith in America" 5: 90–99.

Putnam, R. D. 1993. *Making Democracy Work*. Princeton, NJ: Princeton University Press.

———. 1995. "Bowling Alone: Democracy in America." Nobel Lecture, Stockholm.

Quirk, P. J. 1992. "Structures and Performance: An Evaluation." In *The Post-Reform Congress*, edited by R. H. Davidson. New York: St. Martin's Press.

Rainey, H. G., R. W. Backoff, and C. H. Levine. 1976. "Comparing Public and Private Organizations." *Public Administration Review* 36: 233–44.

Rawls, J. 1971. *A Theory of Justice*. Cambridge, MA: Harvard University Press.

Rhodes, R. A. W. 1997. *Understanding Governance*. Buckingham, UK: Open University Press.

Richardson, J. J., and A. G. Jordan. 1994. Governing Under Pressure. Oxford: Martin Robertson.

Riley, D. D. 1987. *Controlling the Federal Bureaucracy*. Philadelphia: Temple University Press.

Rockman, B. A. 1990. "Entrepreneurs in the Political Marketplace: The Constitution and the Development of the President." Paper presented at the Annual Meeting of the American Political Science Aassociation, San Francisco.

———. 1991. "The Leadership Question: Is There an Answer?" In *Executive Leadership in Anglo–American Systems*, edited by C. Campbell and M. J. Wyszomirski. Pittsburgh: University of Pittsburgh Press.

Roeberg, V. 1991. "Politikkutforming gjennom personellrekruttering?" Thesis. Department of Political Science, University of Oslo.

Rogowski, R. 1972. *Rational Legitimacy*. Princeton, NJ: Princeton University Press.

Rohde, D. W. 1989. "'Something's Happening Here: What It Is Ain't Exactly Clear.' Southern Democrats in the House of Representatives." In *Home Style and the Washington Work: Studies of Congressional Politics*, edited by M. P. Fiorina and D. W. Rohde. Ann Arbor: University of Michigan Press.

Rohr, J. A. 1987. "The Administrative State and the Constitutional Principle." In *A Centennial History of the American Administrative State*, edited by R. C. Chandler. New York: Free Press.

Rokkan, S. 1966a. "Comparative Cross-National Research: The Context of Current Efforts." In *Comparing Nations: The Use of Quantitative Data in Cross-National Research*, edited by R. L. Merritt and S. Rokkan. New Haven, CT: Yale University Press.

———. 1970. *Citizens, Elections, Parties*. Oslo: Norwegian University Press.

Rommetvedt, H. 1994. "Personellressurser, aktivitetsnivå og innfly-telsesmuligheter i et Storting i vekst." Paper Presented at the National Conference in Political Science, Geilo.

Rose, R. 1976. *The Problem of Part Government*. London: Macmillan.

———. 1980. "Government Against Sub-Government." In *Presidents and Prime Ministers*, edited by R. Rose and E. N. Suleiman. New York: Holmes and Meier.

Sabatier, P. A. 1989. "A Policy-Advocacy Coalition Model of Policy Change and the Role of Policy Learning Therein." *Policy Sciences* 21: 129–68.

Sabine, G. H., and T. L. Thorson. 1973. *A History of Political Theory*. Homewood, IL: Dryden Press.

Salamon, L. B. 1981. "The Question of Goals." In *Federal Reorganization*, edited by P. Szanton. Chatham, NJ: Chatham House.

Sartori, G. 1970. "Concept Misinformation in Comparative Analysis." *American Political Science Review* 64: 1033–53.

———. 1991. "Comparing and Miscomparing." *Journal of Theoretical Politics* 3: 243– 57.

Savoie, D. J. 1995. "Central Agencies: Looking Back." Unpublished Paper, Canadian Centre for Management Development, Ottawa.

———. Forthcoming. *Governing from the Centre*. Toronto: University of Toronto Press.

Sbragia, A. M. 1996. *Debt Wish*. Pittsburgh: University of Pittsburgh Press.

Schattschneider, E. E. 1960. *The Semisovereign People*. Chicago: Dryden.

Schlesinger, A. M. 1992. *The Disuniting of America*. New York: W. W. Norton.

Schoenbrod, D. 1993. *Power Without Responsibility: How Congress Abuses the People Through Delegation*. New Haven, CT: Yale University Press.

Schon, D., and M. Rein. 1994. *Frame Reflection: Resolving Intractable Policy Disputes*. New York: Basic Books.

Schulman, P. 1991. "The Politics of Ideational Policy." *Journal of Politics* 53: 114–45.

Schuman, D., and D. W. Olufs III. 1988. *Public Administration in the United States*. Lexington, MA: Heath.

Scott, W. R. 1992. *Organizations: Rational, Natural and Open Systems*. Englewood Cliffs, NJ: Prentice-Hall.

Seidman, H. 1980. *Politics, Position, and Power: The Dynamics of Federal Organization*, 3d ed. New York: Oxford University Press.

Seidman, H., and R. Gilmour. 1986. *Politics, Power and Position*, 4th ed. New York: Oxford University Press.

Seigfried, A. 1940. *France: A Study in Nationality*. Cambridge, MA: Harvard University Press.

Seip, J. A. 1963. *Fra embetsmannsstat til ettpartistat og andre essays*. Oslo: Norwegian University Press.

Selle, P. 1980. "Stortingskomiteane 1814–1981," report no. 89. *The Study of the Distribution of Power in Norway*. Bergen.

Selznick, P. 1949. *TVA and the Grass Roots*. Berkeley: University of California Press.

———. 1957. *Leadership in Administration*. New York: Harper and Row.

Shepsle, K. A. 1988. "Representation and Governance: The Great Legislative Trade-off." *Political Science Quarterly* 103: 461–83.

Shugart, M. S., and J. M. Carey. 1992. *Presidents and Assemblies: Constitutional Design and Electoral Dynamics*. Cambridge: Cambridge University Press.

Simmel, G. 1964. *Power and Conflict*. New York: Free Press.

Simon, H. A. 1976. *Administrative Behavior*, 3d ed. New York: Free Press.

Sinclair, B. 1993. "House Majority Party Leadership in an Era of Divided Control." In *Congress Reconsidered*, 5th ed., edited by L. C. Dodd and B. I. Oppenheimer. Washington, DC: CQ Press.

Skocpol, T. 1979. *States and Social Revolutions*. Cambridge: Cambridge University Press.

Smelser, N. J. 1973. "The Methodology of Comparative Analysis." In *Comparative Research Methods*, edited by D. Warwick and S. Osherson. Englewood Cliffs, NJ: Prentice-Hall.

Smith, E. 1993. *Høyesterett og folkestyret*. Oslo: Norwegian University Press.

Smith, M. J., D. Marsh, and D. Richards. 1993. "Central Government Departments and the Policy Process." *Public Administration* 71: 67–94.

Smith, S. S. 1992. "The Senate in the Postreform Era." In *The Postreform Congress*, edited by R. H. Davidson. New York: St. Martin's Press.

Spicer, M. W., and S. B. Lundstet. 1976. "Understanding Tax Evasion." *Public Finance* 31: 295–305.

Street, J. 1994. "Review Article: Political Culture—From Civil Culture to Mass Culture." *British Journal of Political Science* 24: 95–114.

Strom, K. 1990. *Minority Government and Majority Rule*. Cambridge: Cambridge University Press.

Sullivan, T. 1987. "Presidental Leadership in Congress: Securing Commitsments." In *Congress: Structure and Policy*, edited by M. D. McCubbins and T. Sullivan. Cambridge: Cambridge University Press.

Sundquist, J. L. 1981. *The Decline and Resurgence of Congress*. Washington, DC: Brookings Institution.

———. 1987. "Congress as Public Administrator." In *A Centennial History of the American Administrative State*, edited by R. C. Chandler. New York: Free Press.

———. 1988. "Needed: A Political Theory for the New Era of Coalition Government in the United States." *Political Science Quarterly* 103: 613–35.

———. 1992. *Constitutional Reform and Effective Government*, rev. ed. Washington, DC: Brookings Institution.

Sunstein, C. R. 1990. *After the Rights Revolution*. Cambridge, MA: Harvard University Press.

Szanton, P. 1981. *Federal Reorganization*. Chatham, NJ: Chatham House.

Thiebault, J.-L. 1991a. "The Social Background of Western-European Cabinet Ministers." In *The Profession of Government Minister in Western Europe*, edited by J. Blondel and J.-L. Thiebault. Basingstoke, UK: Macmillan.

———. 1991b. "Local and Regional Politics and Cabinet Membership." In *The Profession of Government Minister in Western Europe*, edited by J. Blondel and J.-L. Thiebault. Basingstoke, UK: Macmillan.

Thurber, J. A. 1988. "The Concequences of Budget Reform for Congressional-Presidental Relations." In *Congress and the Presidency: Invitation to Struggle. The Annals of the American Academy of Political and Social Science*, edited by R. H. Davidson. Newbury Park, CA: Sage.

———. 1991. "Introduction: The Roots of Divided Government." In *Divided Democracy: Cooperation and Conflict Between the President and Congress*, edited by J. A. Thurber. Washington, DC: CQ Press.

Van Riper, P. P. 1987a. *History of the United States Civil Service*. Westport, CT: Greenwood.

———. 1987b. "The American Administrative State: Wilson and the Founders." In *A Centennial History of the American Administrative State*, edited by R. C. Chandler. New York: Free Press.

Veld, R. 1992. *Autopoesis and Configuration Theory*. Dordrecht: Kluwer.

Vogel, J. 1974. "Taxation and Public Opinion in Sweden: An Interpretation of Recent Survey Data." *National Tax Journal* 27: 499–513.

Walcott, C., and K. M. Hult. 1987. "Organizing the White House: Structure, Environment, and Organizational Governance." *American Journal of Political Science* 31: 109–25.

Walsh, K., and J. Stewart. 1992. "Change in the Management of Public Services." *Public Administration* 70: 499–518.

Watson, R. A. 1988. "The President's Veto Power." In *Congress and the Presidency: Invitation to Struggle. The Annals of the American Academy of Political and Social Science*, edited by R. H. Davidson. Newbury Park, CA: Sage.

Weaver, R. K. 1989. *Automatic Government*. Washington, DC: Brookings Institution.

Weaver, R. K., and B. A. Rockman. 1993. "Assessing the Effects of Institutions." In *Do Institutions Matter? Government Capabilities in the United States and Abroad*, edited by R. K. Weaver and B. A. Rockman. Washington, DC: Brookings Institution.

Weber, M. 1970. "Bureaucracy." In *From Max Weber*, edited by H. H. Gerth and C. W. Mills. London: Routledge and Kegan Paul.

Weimer, D. 1995. *Institutional Design*. Dordrecht: Kluwer.

West, D. M. 1988. "Gramm-Rudman-Hollins and the Politics of Deficit Reduction." In *Congress and the Presidency: Invitation to Struggle. The Annals of the American Academy of Political and Social Science*, edited by R. H. Davidson. Newbury Park, CA: Sage.

White, L. D. 1948. *The Federalist Era: A Study in Administrative History*. New York: Macmillan.

———. 1958. *The Republican Era, 1869–1901: A Study in Administrative History.* New York: Macmillan.

Wildavsky, A. 1987. *The New Politics of the Budgetary Process*. Boston: Little, Brown.

Wilson, J. Q. 1987. "Does the Separation of Powers Still Work?" *Public Interest* 86 (winter): 36–52.

Wilson, R. W. 1992. *Compliance Ideologies: Rethinking Political Cultures*. Cambridge: Cambridge University Press.

Wilson, W. 1887. "The Study of Administration." *Political Science Quarterly* 2: 197–222.

Wolin, S. S. 1960. *Politics and Vision. Continuity and Innovation in Western Political Thought*. Boston: Little, Brown.

Wyszomirski, M. J. 1991. "The Discontinuous Institutional Presidency" In *Executive Leadership in Anglo–American Systems*, edited by C. Campbell and M. J. Wyszomirski. Pittsburgh: University of Pittsburgh Press.

Index

About the Authors

TOM CHRISTENSEN is Professor of Political Science and Public Administration at the University of Oslo. He has published extensively on the public sector in Norway, as well as a number of articles on comparative public administration.

B. GUY PETERS is Maurice Falk Professor of Government at the University of Pittsburgh and Research Fellow of the Department of Government, University of Strathclyde. His recent publications include *Institutional Theory in Political Science* and *Comparative Politics: Theory and Method.*